Appropriate Technology
Technology with a Human Face

P. D. Dunn

SCHOCKEN BOOKS · NEW YORK

Prelims ii

First published by SCHOCKEN BOOKS 1979

Copyright © P. D. Dunn 1978

Library of Congress Cataloging in Publication Data

Dunn, Peter, 1927–
    Appropriate technology.

    Includes index.
    1. Technology.   2. Underdeveloped areas —
    Technology.   I. Title
    T49.5.D86   1979   609'.172'4   78–18246

Printed in Hong Kong

Contents

Foreword by G. McRobie   v
Introduction   vii
Acknowledgements   xi

  1 An Introduction to Appropriate Technology   1
  2 The State of the World — The Views of Economists
      and the Predictions of Futurologists   18
  3 Intermediate Technology and Appropriate
      Technology   40
  4 Food, Agriculture and Agricultural Engineering   55
  5 Water and Health   85
  6 Energy   111
  7 Services — Medicine, Building, Transport   133
  8 Small Industries in Rural Areas   146
  9 Education, Training, Research and Development   158
 10 Getting Started   174

Appendices
    I Conversion Factors   179
   II Gross Domestic Product, Gross National Product
       and National Income   181
  III Project Proposals by I.T.D.G. University Liaison
       Unit   183
  IV Water-related Diseases   185
   V Energy, Work and Power   190
  VI Some Educational Statistics   192
 VII Official Aid   198
VIII Useful Addresses   200

References   203
Further Reading   209
Index   215

Foreword

In the summer of 1965, a group of people in London formed the Intermediate Technology Development Group. They had in common a variety of overseas experience; the conviction that 'development' means first and foremost the development of people; and the knowledge that aid and development as currently practised was bypassing the majority of poor people in the developing countries.

Membership and support grew rapidly, and soon the Group had a nucleus of staff and had started work on filling the 'knowledge gap' about self-help technologies: technologies that are relatively small, simple, and capital-saving, and therefore more appropriate to the needs and resources of the majority of poor people in the rural and small-town areas of the developing countries.

These modest efforts nearly thirteen years ago have since grown into a world-wide network of people and organisations concerned with the development and application of Appropriate Technologies. There are now many groups—representing various combinations of professional people, universities, industry and government—engaged on such tasks as scaling down large-scale technology to bring it within the reach of rural people, up-grading traditional crafts to make them more productive and versatile, and inventing new tools and equipment with which the poor can work themselves out of poverty.

Work on Appropriate Technology is going ahead in many countries in Africa, in the Indian sub-continent, the Far East, and Latin America; and as I write more are in process of formation. Many of these—indeed most of them—were started by small groups supported by voluntary agencies throughout the world. Today, this spearheading voluntary effort is being backed up by government and international support. Work programmes in Appropriate Technology are being built into the official aid programmes of the British, American, Netherlands, German and Scandinavian aid programmes, and a growing number of

U.N. agencies. So we are at least on the threshold of a significant shift in the allocation of aid and development resources, away from the heedless promotion of technologies that make the rich richer, and towards the development of technologies that can enable the rural and urban poor to earn a decent living.

One of the pioneers in the work of developing Appropriate Technologies, and helping others to do so, is the author of this book. Professor Dunn was one of the first distinguished scholars and academics to become an active supporter of the Intermediate Technology Development Group. As Chairman of the Group's Power Panel he has systematically built up a work programme on small-scale energy sources; and he has also assisted several universities in developing countries on the launching of their own research and development work on Appropriate Technologies based on renewable energy sources, and he has advised on the setting up of small industry. His work—and this book—demonstrate beyond question that the discovery of low-cost, small-scale, sustainable technologies offers a whole new range of challenges and opportunities to scientists and engineers—and especially to the new generation of technologists coming up in the developing countries.

GEORGE McROBIE
I.T.D.G., London
December 1977

Introduction

This book is about a self-help approach to development, a method variously described as Intermediate Technology or Appropriate Technology. The term 'intermediate technology' was first coined by Dr E. F. Schumacher to express his views on how development might be achieved, and his ideas have been further extended and elaborated in his book *Small is Beautiful*. Whilst Intermediate Technology is concerned with techniques, Appropriate Technology covers, in addition, the social aspects of development. Dr Schumacher's work and writings have been responsible for much of the recent upsurge of interest in the application of appropriate technology techniques in the developing countries. Its general aims are summarised in the quotation at the head of chapter 1.

Development has been going on for many years and a large number of highly competent and experienced professionals have been, and still are, engaged in it. Quite often these people have been practising Appropriate Technology but not under that name, rather like the man in the Molière play who was surprised to find that he had been speaking prose for the last forty years. What then, if anything, is new about Appropriate Technology? I think that the principal feature about Appropriate Technology which is new is that it offers a complete package solution to the development problems of a particular community rather than a piecemeal list of particular solutions. This package is appropriate to the local skills and other resources and offers the prospect of continuous development in the future. An important requirement of the field workers is information on the availability of low-cost equipment, information on techniques, and designs for equipment for local manufacture. Over the last ten years or so a number of organisations have been set up in different parts of the world to satisfy these needs. One of the earliest of these organisations was the Intermediate Technology Development Group Ltd (I.T.D.G.), started in 1965 by Dr Schumacher and two of his colleagues, George McRobie and Mrs Julia Porter.

This book attempts to provide a systematic and comprehensive general treatment of the subject. It is always valuable to see the place of one's work in the general social context and to have some idea of long-term future trends. Not only is this important in order to have some feeling for priorities but such knowledge also makes life so much more interesting. For these reasons I have outlined some of the more important views of the development economists and also those of the futurologists, and hope that this introduction will stimulate more reading in these fascinating fields. I am not an economist, and should I inadvertently

9 King Street, I.T.D.G. Headquarters
(Photo Peter Fraenkel)

have misrepresented their, or any other specialist's, views I apologise; this of course is a risk one takes in writing on a broad range of topics. This book is concerned essentially with poverty and methods by which it might be alleviated, and inevitably this has political implications which often in a particular situation cannot be ignored. Social, political and economic circumstances differ very widely and views on them differ even more; so, whilst recognising its local importance, no political view has been expressed here. Most of the book is devoted to the actual

practice of Appropriate Technology. The subjects treated include food and agriculture, water and health, energy, medical
services, building and other services, small industry, education
and research. It is intended to provide background information
and both to supplement and introduce the specialised textbooks
and handbooks.

Since it is expected that the specialist knowledge of readers will
vary very widely, the standard has been selected to be at a level
suitable for the intelligent general reader. Mathematics is not
used and specialist terms have been avoided as far as possible.
Where they have been required, for example in distinguishing
between energy and power, or in referring to G.N.P., they have
been explained in appendices.

Statistical information is also presented in appendices. Short
lists of general reading have been given at the end of the book.
Bacon once wrote that 'some books are to be tasted, others to be
swallowed, and some few to be chewed and digested'. Most of
the books listed under general reading fall under the latter
category and I hope will give the reader as much pleasure as they
have given me.

Appropriate Technology can, I believe, make a significant
contribution to solving one of the greatest human problems, that
of poverty. There is no doubt of the worthwhile nature of
working in this field; it also happens to be interesting, satisfying
and great fun. I am not saying of course that this work is never
distressing, exasperating, frustrating and hard—it often is, but
so are all human activities from time to time. However, the
positive returns far outweigh these negative aspects. For those
readers who would like to learn more about it, I have suggested
some ways in which they can do this, in chapter 10. I hope that a
few might actually want to join us in the work.

Acknowledgements

I would like to thank colleagues for their help and advice (which I have not always taken), and in particular wish to mention Colin Allsebrook, Richard Burton, Irwin Friedeman, Ian Gibb, Joe Levy, John Powell, Reg Scott and Martin Upton, who have read and commented on individual chapters, and in some cases the whole book, and to Charles Preston for his help in preparing diagrams.

I also wish to express my thanks to a number of authors and organisations for generously allowing me to reproduce their photographs and diagrams. Credits have been given where the material appears in the book. I.T.D.G. and I.T.D.G. colleagues Peter Fraenkel, Vernon Littlewood, Jane Landymore and Frank Solomon have been especially helpful.

1

An Introduction to Appropriate Technology

If you want to go places, start from where you are.
If you are poor, start with something cheap.
If you are uneducated, start with something relatively simple.
If you live in a poor environment, and poverty makes markets small, start with something small.
If you are unemployed, start using your labour power; because any productive use of it is better than letting it lie idle.
In other words, we must learn to recognise the boundaries of poverty. A project that does not fit, educationally and organizationally, into the environment, will be an economic failure and a cause of disruption.

E. F. Schumacher

The population of the world is distributed over very varied regions differing markedly in climate and environment. There are also large ethnic, social and cultural differences between the various communities and their economic conditions. Roughly one-quarter of the world's population is generally regarded as developed and the remaining three-quarters is described variously as developing, undeveloped, Third World, emerging, or low income. Whatever word we use to describe these countries, the basic fact about them is that they are regions in which most of the people have a low, and sometimes a very low, standard of life.

Great efforts have been made over the past two or three decades to close the gap between the developed and the developing countries and there have been some striking successes, but the general picture in the developing countries is disappointing: average growth has been slow. The result of this has been a widening of the gap between the developed and the developing countries. One method of measuring this gap is in terms of Gross National Product (G.N.P.) per head. In these terms the average

1

citizen in a developed country is now 20 times better off than his counterpart in a developing country. (Figure 1.1 illustrates this in terms of National Income per head which is closely related to G.N.P. per head.) G.N.P. may not be the best criteria by which to compare relative standards, and other, non–monetary, indicators are discussed later; however, all these indicators or criteria show the relative stagnation of the less developed countries and the increasing gap between them and the developed nations.

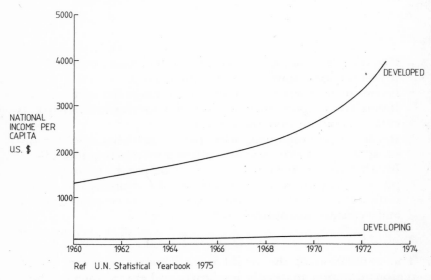

Ref   U.N. Statistical Yearbook 1975

FIG. 1.1   Growth of National Income per capita for the developing and developed countries

The recent improvement in mass communications, and in particular the availability of the now ubiquitous transistor radio, has brought a new awareness of the differences in standards of living and an increasing impatience with the slow rate of change. Expectations are outrunning achievements.

Whilst the greater part of the world's population lives under conditions of extreme poverty, there does not seem to be an imminent danger of famine on a world scale, though local areas are vulnerable to crop failure. Food production has increased roughly in step with population over the last two decades. Nevertheless diet is often of poor quality, and also inadequate in quantity. Seasonal food shortages are quite common, and children, nursing mothers, and the sick are particularly at risk.

The main impetus of development programmes so far has

been directed towards the construction of large capital plant—steelworks, chemical works, roads, dams, airports and power stations. These are undoubtedly an essential component of development but are by no means the whole story. One result of such programmes has been the growth of a dual society, islands of development, usually in the towns, populated by a highly educated élite but surrounded by a stagnant rural sea. It is in the rural areas where most of the population live, typically 70–80 per cent, and here the pattern is not of growth but of decay, lack of development, mounting unemployment and drift to the towns.

The basic problem is that of combating poverty, and particularly poverty in the rural areas. It is often suggested that the choice facing developing countries is between retaining traditional practices and the discontinuous jump to a modern technological society. Intermediate Technology was proposed by E. F. Schumacher as a new route to development, a self-help approach. Appropriate Technology is the preferred, more comprehensive, term and its aims are summarised in the quotation at the beginning of this chapter.

Appropriate Technology and how it started

Dr Schumacher first presented his ideas under the name Intermediate Technology at a seminar on 'Technologies for Small Industries in Rural Areas' organised by the Indian Planning Commission at Hyderabad in March 1964. An expanded and generalised account of Dr Schumacher's views was later published, under the title *Small is Beautiful*,[1] an outstanding and influential work.

The Appropriate Technology method attempts to recognise the potential of a particular community and tries to help it to develop in a gradual way. This development is based on local resources and progressively builds up the skills of the community. It is essentially rural-based rather than urban-based as is the Western technology. One important feature is the emphasis on the creation of work places. The work place in the developed countries will cost, typically, £2,000 to £3,000 in terms of capital investment. This is clearly an impossibly high figure in a developing country, which will have little capital and even less foreign exchange with which to import equipment. Such a country will, however, have a plentiful supply of labour even though this will be unskilled and unfamiliar with industrial practice. Appropriate Technology can be regarded as centring around job creation for these people.

It is important to understand that Appropriate Technology is not only concerned with small industry, but covers all aspects of community development. It is a complete systems approach to development and is self-adaptive and dynamic, in that as people become wealthier and more skilled they can both afford and also use more expensive equipment.

In 1965, together with George McRobie and Mrs Julia Porter, Dr Schumacher set up the Intermediate Technology Development Group Ltd (I.T.D.G.), a charitable trust. The object of I.T.D.G. was to promote the Intermediate or Appropriate Technology concept and, as one of its principal activities, it was intended to assist in the collection and dissemination of data on simple low-cost technologies.

Other associated developments

People have been practising Intermediate or Appropriate Technology since the beginning of civilisation, though not by these names.[2] A more recent example is the establishing of the industrial base in the United States which started from small beginnings in the early part of the nineteenth century. Small industry and craft skills were built up at this time in order to provide alternatives to the manufactured goods whose supply from Britain was interrupted by the Napoleonic Wars.

At the commencement of the present century, in India, Gandhi stressed the importance of encouraging the development of village industry. The charka (spinning wheel) was chosen as the symbol for this movement, and (after independence) was in fact proposed, unsuccessfully, as the symbol to appear on the national flag. Gandhi initiated the systematic study of village technology and a large number of village industrial units were formed. There are now several thousand of these units in India.

In China, Mao and his colleagues were associated with the country areas not the town, and in fact the social revolution was brought about by the rural peasants rather than the urban proletariat of the Marxist teachings. The Chinese tended to concentrate on the build up of heavy industry until the break with Soviet Russia in 1960; after this time they reversed their policy and shifted to decentralisation and the encouragement of self-sufficiency at a local level. This latter policy is summed up by the well-known Mao quotation 'walking on two legs'. Very impressive advances of an Appropriate Technology character have been reported.

What is new is the formulation of the Appropriate Technology concept as a total development package for the solution of a community development problem.

Following Dr Schumacher's initiative several Appropriate Technology centres have been set up in the developing world, including centres in Kumasi, Ghana, and Lahore, India. The latter may be regarded as continuing the work started by Gandhi. In addition there are now a number of organisations of a similar nature to I.T.D.G. in the developed world. These include V.I.T.A. in the U.S.A., Brace in Canada, and T.O.O.L. in Holland.

The aims of development—what A.T. tries to do

The principal aims of development are

    (1)   To improve the quality of life of the people.
    (2)   To maximise the use of renewable resources.
    (3)   To create work places where the people now live.

The solutions chosen should satisfy the following criteria

    (1)   Employ local skills.
    (2)   Employ local material resources.
    (3)   Employ local financial resources.
    (4)   Be compatible with local culture and practices.
    (5)   Satisfy local wishes and needs.

What then are the basic requirements of a community? They may be listed under the following broad headings

    (1)   Food.
    (2)   Water.
    (3)   Clothing.
    (4)   Shelter.
    (5)   Health care, hygiene, and sanitation.
    (6)   Education and training.

The latter is essential if the community is to develop. In addition, though it does not appear on this list, man requires a source of energy for domestic purposes and to supplement his own muscle power. The particular requirements of a community will depend on its stage of development. Identifying these needs is a complex problem to which there is no simple solution. Different solutions will be required for different areas. Some of the problems are of a political nature, others social and cultural. Land tenure is a serious difficulty in many parts of the developing world. However, even when the political climate and the social will is there, it is still necessary to identify appropriate solutions and to arrange for their implementation.

In a particular community there will be traditional agricultural practices and in many cases an indigenous artisan industry, it is however a largely static situation, and what is required is the fostering in the people of a development attitude, a belief that progress can be made by self-help methods.

It should be remembered that a 'subsistence economy' means just what it says. Such people have, over generations, evolved a way of life that is just adequate. They have very little margin available for risk taking. Their aims are not to maximise return but to avoid disaster. Such communities are sometimes criticised for their conservatism and unwillingness to depart from traditional practice, whereas in fact, their preference for the well-tried is often essential for survival.

It is sometimes suggested that Appropriate Technology is based on the superimposition of an outside technology on a community. This is not in fact so; A.T. aims instead to encourage and point the way to change by using one's own resources. This is the way of 'gradualness' as opposed to the 'big push' advocated by some development economists (though not all of them; Singer and others have pointed out that the capital and material resources required for such a jump are just not available). It is also questionable whether such a discontinuity, even if possible, would be socially desirable because of the inevitable consequent disruption in community practices and values. The acceptance of change can often be initiated by the provision of a single piece of equipment of community value—for example, a pump. Once this acceptance has been achieved a broad development programme covering all the community needs can be initiated.

Some examples of Appropriate Technology equipment

In this section several specific examples are described, covering most of the items listed in the previous section. There are very

FIG. 1.2   Cretan–type windmill in Northern Ethiopia *(Photo Peter Fraenkel)*

wide differences between individual areas of the developing world, ranging from subsistence farming to large industries in the towns. Thus a solution which is appropriate in one situation may be unsuitable in another, either because it is too complex or insufficiently advanced. The following examples have been selected to include a range of complexity from the very simple to the relatively advanced. Figure 1.2 shows a canvas–sailed Cretan-type windmill introduced to a village community in Northern Ethiopia, which could serve as the introduction of A.T. equipment to this area.[3] Such a windmill will power a small pump and irrigate about an acre of land. The increased availability of water both improves the yield from traditional crops and also allows the introduction of new crops, for example, in the setting up of kitchen gardens. The result is an improvement in quantity and quality of the diet. The windmills were of standard Cretan-type design and this is an example of how an established technology in one region can be transplanted to another. The mills shown were constructed in the workshop of the local mission but could equally well have been made in the village. A major advantage of such local manufacture is the availability of maintenance skills and replacement parts. Most important is the educational content and the change in attitudes resulting from the introduction of this equipment. In passing, one should note the need to consider social and related factors: in the case of the first canvas-

FIG. 1.3   I.T.D.G. water catchment tank *(Photo I.T.D.G.)*

FIG 1.4   I.T.D.G. water
catchment tank (beehive-
type) *(Photo I.T.D.G.)*

sailed windmills, the sails proved to be an irresistible attraction to
the village girls for use as skirt material until replaced by a
material less suited to the latter application.

A second example is the introduction of rainwater catchment
tanks in backward areas of Botswana. The rainfall in Botswana
is seasonal and there is a need for some means of storing water for
use in the dry season. The water is required for the normal
domestic purposes, drinking, washing and cooking, and is also
needed for watering cattle and irrigation of crops. The idea of
storing rainwater is not new but it can be made both cheap and
efficient by the use of modern materials.[4] Figures 1.3 and 1.4
show the general form of construction. A 10,000 gallon capacity
tank can be constructed using 120 man hours of unskilled labour
and at a cost for plastic materials of £12.50. Basic material is thin
polythene or butyl rubber sheet, together with tubes of around
three inches diameter of the same material. An excavated pit is
lined with alternate layers of mud slurry and the thin sheeting.
An insecticide is mixed with the mud. The walls and floor have a
final lining made from 'sausages' constructed from the polythene
tube which is filled with a mixture of cement and sand. Because
of the method of construction, the cement required is only a
quarter of that needed for normal cement blocks. An apron of
sheeting is arranged to catch rainwater and allow it to flow into

the tank and the tank is completed by the addition of a lid. A sand filter can be incorporated by constructing beehive-shaped domes from further sausages and covering the outside of the domes with sand. It is important to fence in the tank and apron otherwise goats will damage the latter (figure 1.5). The combination of insatiable curiosity, sharp cloven hooves and cast iron digestive systems, makes these animals a universal scourge where equipment is concerned.

FIG. 1.5   Goats damaging catchment tank *(Photo I. T.D.G.)*

The design of the water tank was developed by I.T.D.G. who, having produced a prototype, were faced with the problem of disseminating information amongst the villages. The solution adopted is interesting and could be more widely used. A school in Botswana was hired for two weeks during the holiday period. Some forty teachers from outlying villages were invited to join a course for which no fee was charged. The course consisted of constructing a tank and an associated school kitchen garden. At the end of the course the teachers were given a pack of plastic material for making a new tank. It was hoped that on their return to their own school they would construct a school tank using the children as labour. In this way not only was there a local example of a tank for villagers to see but the know-how had also been transferred through the children. Some eleven tanks were actually constructed in primary schools and kitchen gardens of about 200 square yards irrigated throughout the dry season. These gardens produced sufficient green vegetables to provide ninety children with vegetables for two meals a week. In this example the development of a new material has been used to give a new solution, and a useful method of spreading the information has been identified.

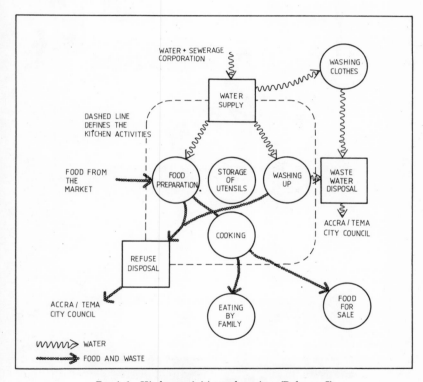

FIG. 1.6   Kitchen activities and services *(Reference 5)*

An example of Appropriate Technology applied to housing is found in Tema, a new city-port in Ghana.[5] The house was designed in the early 1970s by a cooperative including tenants, builders and planners. The basic unit is a ten-foot square kitchen and a courtyard, around which up to four additional rooms may be added. Considerable thought was put into analysing the basic activities when planning the kitchen (figure 1.6). The design finally adopted is shown in figures 1.7 and 1.8. Locally available materials were used in the construction of the building. The

FIG. 1.7　Interior of Tema house *(Photo I.T.D.G.)*

FIG. 1.8　Exterior of Tema house *(Photo I.T.D.G.)*

walls were built from blocks made from cement and sand, the roof cladding and kitchen shutter were made from corrugated aluminium sheeting. Second-grade African hardwood was used for the structural timber and the boards from packing cases for the ceiling cladding. Locks and other fastenings were made by local blacksmiths. Since the first prototype houses have been occupied the tenants have made suggestions for improvements in later houses.

Not all Appropriate Technology involves actual equipment; techniques are equally important particularly in education and medical care. Freire reports on a new method of teaching adult illiterates to read which he has developed in South America and which has proved to be highly successful. Briefly, the method adopted is to select some trisyllabic word having an emotive political or social significance for the group—for example, FAVELA meaning 'slum'. The significance of the word is first discussed by the group and then it is written down and broken up into its basic syllables, FA VE LA. The family of the first syllable is then derived as an introduction to the vowels, FA, FE, FI, FO, FU. This process is repeated for the remaining two syllables. The members of the group are then encouraged to create further simple words using these basic elements. Typical suggestions include FAVO, LUVA, LI, VALE, VIVE, VIVO, etc. This method works well with Spanish but would not necessarily be suited to languages of a different structure; however, we should remember that there are very many illiterate Spanish-speaking people, particularly in South America, so the potential of the method is considerable. Freire's method is outlined in his book, *Cultural Action for Freedom*.[6] In it he quotes a rather touching remark made to him by one peasant: 'Before, letters seemed like little puppets. Today they say something to me, and I can make them talk.' There is currently considerable interest in the development of new methods of medical care in order to improve the health in the rural areas. Notable examples include the Chinese 'barefoot doctor' scheme and a similar scheme for medical auxiliaries in Tanzania; these are described in chapter 7.

By providing designs which enable equipment to be made locally, work places are created and skills increased. Figure 1.9 shows examples of hospital equipment which have been designed and built in Zaria in Northern Nigeria. The cost of this equipment is substantially lower than, and typically half the cost of, imported equivalents; in addition it has a longer life since it has been designed for local conditions—rough concrete floors

FIG. 1.9   Invalid carriages constructed in Zaria, Nigeria *(Photo I.T.D.G.)*

rather than smooth plastic tiles — and can readily be repaired. No foreign currency is required.

During a visit to Zambia in 1969 Dr Schumacher found a requirement for a means of packaging eggs from farms to central depots. The conventional solution is to use trays made from paper pulp and to pack these trays in wooden crates (figure 1.10a). In Zambia both the trays and the crates were imported. The problem was tackled by Thomas Kuby, a graduate of the Royal School of Art, as part of his M.Sc. course work. His design (figure 1.10b) has the merit of an interlocking construc-

FIG. 1.10   (a)   Conventional egg trays   FIG. 1.10   (b)   Interlocking egg trays

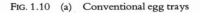

(Photo I.T.D.G.)

tion which enables the stacks of trays to be self-supporting and eliminates the need for a packing case. The stack is secured by means of two lengths of rope. Commercial egg trays are made from paper pulp which is compressed in a mould and later dried. This method of manufacture would have proved quite satisfactory for the present application, but unfortunately it was found that the only available plant was too large. The Zambian requirement was for one million trays a year, whereas the commercial plant had an output of twelve million trays a year; at a cost of £150,000 it was also far too expensive. In addition to having a smaller output, a machine for this work needed to be easy to install, maintain and operate; it must also use local materials such as waste paper, wood, bamboo, or elephant grass. It appeared that there were large supplies of waste paper available for this purpose in Zambia, so this was chosen as the feedstock. Kuby designed a simple pulper and also a press which compressed the low consistency pulp between a set of moulds; the lower mould was fitted with a mesh screen to allow the water to drain away, the draining process being assisted by suction. A prototype was constructed (figure 1.11) and tested in the engineering workshop at Reading University. This machine produced 300,000 trays a year and the cost of the commercial version was only £8,000. A

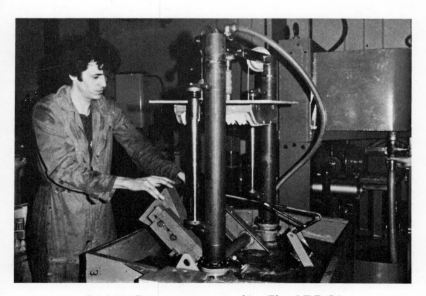

FIG. 1.11   Prototype egg tray machine *(Photo I.T.D.G.)*

Fig. 1.12   Egg tray manufacture in Zambia *(Photo I.T.D.G.)*

simple dryer was also developed to enable the trays to be dried on leaving the press. Four of these machines satisfied the Zambian need (figure 1.12). Since this time further machines have been supplied to other parts of the world. The egg tray machine is a good example of the scaling down of commercial equipment to meet a smaller need, and it does not necessarily lead to an increase in cost of the product.

A labour–intensive brickmaking factory is described in chapter 7.

A cautionary tale

Not all innovations are successful and it is important to assess not only the advantage of a change to the particular individual or firm, but also the total social cost. We will end on a cautionary note by quoting a report by Keith Marsden of I.L.O.[7] on such a development. He describes a small industry in a North African country which was engaged in producing leather sandals for local use. It was suggested that a cheaper product could be obtained by changing from the traditional material, leather, to plastic. The country imported two plastic-moulding injection machines at a cost of $100,000. This highly automated plant operated very successfully, but the net effect was to put 5,000 leather shoemakers out of work and in turn to reduce the incomes of the makers of leather, glue, thread, fabric linings, tacks, dressings, polishes, hand tools, wooden lasts and boxes, all of whose livelihoods were connected with this industry. In their place were just 40 injection moulding operatives. The P.V.C. plastic used in the process had to be imported from abroad, and since local industry did not have the expertise, spares and maintenance were also needed from outside the country. The overall result was a decline in domestic income, a continuing requirement for foreign currency for the import of materials and equipment, and a decline in the living standard in the area.

2

The State of the World — The Views of Economists and the Predictions of Futurologists

> Myself when young did eagerly frequent
> Doctor and Saint and heard great argument
> About it and about: but evermore
> Came out by the same door as in I went.
> *The Rubaiyat of Omar Khayyam*, E. Fitzgerald

Some three-quarters of the world's population, or about 2,800 million people, have a low standard of living; of these, 1,000 million live in conditions of extreme poverty. Most of these people live in the rural areas and are engaged in agriculture or

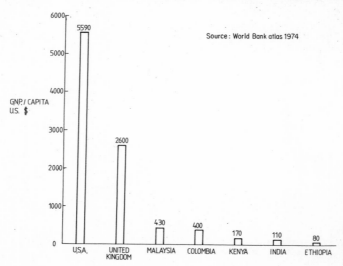

FIG. 2.1   G.N.P. per capita for selected countries (1972 figures)

agriculture-related activities. Figure 2.1 shows the difference in G.N.P. per head between some of the developing and the developed nations. These are of course averages and averages can be misleading, since they can conceal very large individual differences. For example, even in the U.S.A. there is a poverty belt, and similarly in some of the developing countries there are small sectors of considerable wealth. In many of the developing countries the differentials between the very poor sector and that of the very rich are greater than in the industrial nations. We should therefore bear in mind the spectrum of wealth distribution.

Though there is an overlap of the edges of the wealth spectra between the developed and developing nations, nevertheless the fact remains that most of the people of the developing world live under conditions of extreme poverty. G.N.P. is not the most relevant indicator of standard of living—for example, one does not conclude from figure 2.1 that a citizen of the U.S.A. is 50 times better off than a citizen of India. There are also regional differences; an American citizen could not subsist at all in the U.S.A. on the income of an Indian peasant. Non–monetary indicators are more helpful in making comparisons. Figures 2.2, 2.3, 2.4, 2.5, 2.6, 2.7 and 2.8 list some of these—namely, food

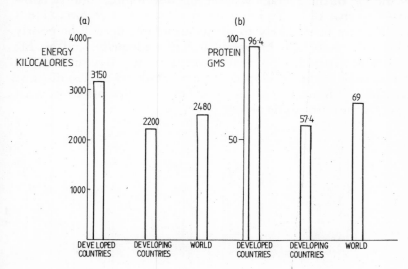

Source : State of Food and Agriculture
(F.A.O. 1975)

FIG. 2.2   Average daily food supplies per capita

20   *Appropriate Technology*

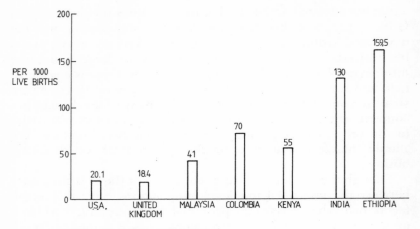

Source: World Tables 1976. World Bank

FIG. 2.3   Infant mortality for selected countries

consumption in terms of total calories and also protein content;
infant mortality per 1,000 births; life expectancy; medical
facilities; literacy; fertiliser use, and energy consumption. These
figures together with G.N.P. per head help to build up a picture
of the life of the average worker in a developing country compared to that of his counterpart in a developed country. He is
seven times more likely to die as an infant, he will be poorly
educated, badly fed, his medical care inadequate, and his life
expectancy perhaps 50 years compared to 70 years for his
counterpart in the developed country. Statistics tend to disguise
realities and figure 2.9 gives a few random illustrations of what
poverty means to some people.

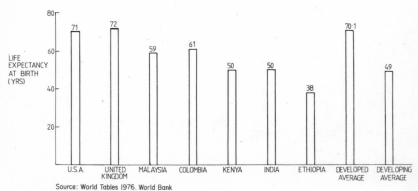

Source: World Tables 1976. World Bank

FIG. 2.4   Life expectancy at birth for selected countries (1970–5 average)

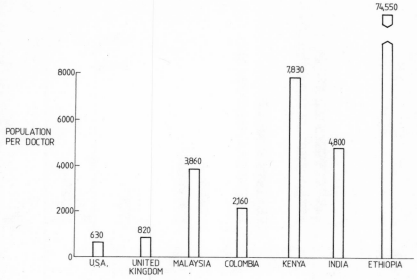

Source: World Tables 1976, World Bank

FIG. 2.5   Population per doctor for selected countries (1970).

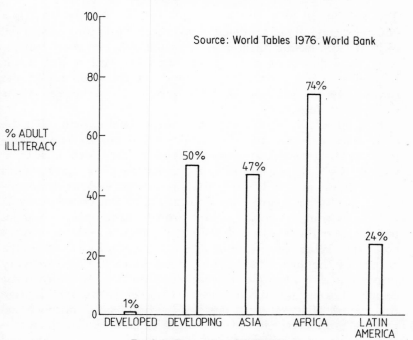

FIG. 2.6   Percentage adult illiteracy

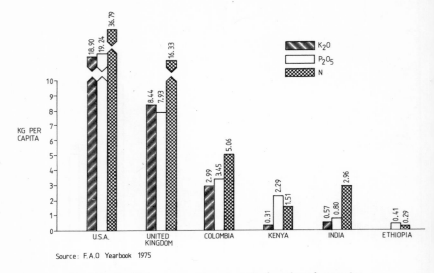

Source: F.A.O Yearbook 1975

FIG. 2.7   Fertiliser consumption per capita for selected countries

Why is the situation like it is, and why does it not improve?

The subject of how the world got into its present state is a complex mass of facts, fallacies and emotive opinions — colonial exploitation, ethnic differences, international company conspiracy, etc. I do not propose to get deeply involved in it, but it is helpful to have a little background on the thinking of the development economists, and also to be aware of the projections of the futurologists.

First of all we should try to define 'development'. To the development economists 'development' seems to be synonymous with 'growth', and growth is measured by the change in G.N.P. G.N.P. is a pretty blunt instrument for dealing with actual living standards, and critics point out various disadvantages. It does not give any indication of the distribution of income within the country, and in most developing countries income distribution is much less equal than it is in the developed countries. When a community is living at a very low standard the nature of the products and other inputs to G.N.P. are very important. For example, the contribution to G.N.P. from the export of luxury goods is less useful to general development than the export of lower price articles having a good home-based market. An increase in G.N.P. might be achieved by the introduction of more efficient, but labour-saving, equipment and

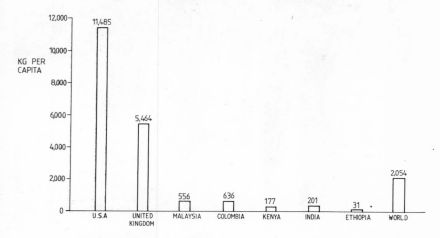

Source : U.N. Statistical Yearbook 1975

FIG. 2.8   Energy consumption per capita for selected countries

result in higher unemployment. Also at the subsistence level immediate and significant improvements in the quality of life may not involve the market economy and will probably not appear in the G.N.P. For example, the introduction of kitchen gardens may result in an important improvement in diet and in health, but will not affect the G.N.P. for perhaps a generation. The same is true of improved health care and education. Nevertheless, in spite of its weaknesses, G.N.P. does provide a crude and convenient indication of development progress.

The world can be seen to be divided into the 'Haves' and the 'Have nots'—that is, the developed and the developing sectors. This state of affairs is sometimes described in the literature as 'dualism' and also exists within the developing countries themselves, where there are small sections of wealth amidst the general poverty. There seems evidence that the gap between the rich and the poor nations and the gaps within the developing countries themselves are not only likely to persist but will actually increase.

An alternative division of the world in terms of broad economic/political differences is adopted in the U.N. literature, which delineates three main groups: the 'developed market' or capitalist economy; the 'centrally planned' or communist economy; and, the third group, the 'developing market' economy, also described as the Third World.

The Third World contains about 90 nations which differ

a

b

c

d

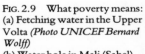

Fig. 2.9 What poverty means:
(a) Fetching water in the Upper
Volta *(Photo UNICEF Bernard
Wolff)*
(b) Water hole in Mali (Sahel)
(Photo UNICEF Diabate)
(c) Water carrying (India) *(Photo
Christian Aid)*
(d) Bangkok slum (Thailand)
(Photo UNICEF Bernard Wolff)
(e) Poor district in Manila
(Philippines) *(Photo UNICEF
Jack Ling)*

e

greatly in state of development, and in social systems, culture and system of government. They range in size from populations of a few million to India with a population in excess of 500 million. Many of the countries have only achieved independence since the end of the Second World War. In addition to this diversity between countries, each country is subject to pressures from its international environment, and any economic plan must take account of these (figure 2.10).

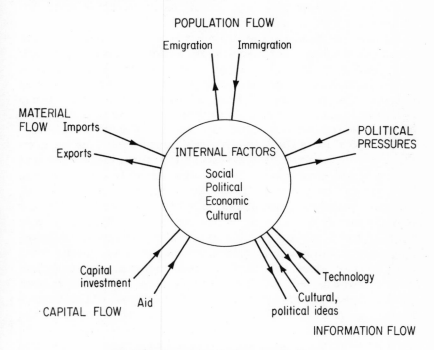

FIG. 2.10   International influences and effects

What development economists do

The role of the development economist is to explain the current situation and lay down policies for the future. There is a wide diversity of theories to be found in the literature. As an interested observer I have noticed a sort of dualism between these theories and am tempted to put forward the general proposition that 'For every economic development theory propounded there will be an economist of equal eminence to that of the proposer who will refute it and probably assert that the reverse is true'.

Predictably the Marxist economists believe that dualism between nations and the stagnation and poverty of the developing world is due to exploitation by the developed countries. They assert that political freedom is insufficient without economic independence, and that the poor nations suffer from a kind of economic colonialism. This is an extension to the international scene of the Marxist philosophy of control of the sources of production by the proletariat. They also recommend comprehensive central planning as an essential requirement for progress.

Equally predictably, the capitalist economists believe in the virtues of the free enterprise system, although it is accepted that all capitalist countries should now operate a modified capitalist system which involves some central planning and government control. The economic systems adopted by the Third World countries range from various combinations of modified capitalism to full communism. One view on why the rate of development is so slow is the 'vicious circle' argument. This suggests that poverty results in low investment; in turn, low investment results in low production, low production in low income, and low income in poverty. There are other vicious circles too. Hunger results in poor physique which in turn leads to low production. A third circle is that of poverty leading to low demand and hence low production. All three circles occur simultaneously and interact. That low income, poverty, hunger, etc., exist is self-evident; that they combine in this manner to inhibit development is less so. The vicious circle argument fails to explain how the present developed countries were able to progress, nor does it explain the recent success of some of the developing countries, and it no longer seems to receive much support from the economists. Myrdal[1] has presented a modified form of the explanation, sometimes referred to as the 'backwash argument', in which he suggests that the existence of the developed nations inhibits the progress of the undeveloped. This theory also has been criticised, notably by Bauer,[2] who asserts that poverty itself is not necessarily incompatible with a rapid rate of growth and development.

Development can be described from the historical point of view. One of the leading exponents of this approach is Rostow,[3] who takes an evolutionary view of development. Rostow suggests that there are defined stages of growth through which society passes from the traditional pattern of life to a condition where economic 'take off' becomes possible, leading in turn to a phase of increased productivity and then to one of high mass consumption. According to Rostow, 'take off' requires a saving rate of 15–20 per cent of G.N.P. Views on the significance of historical development and on its interpretation are contentious and the 'stages of growth' theory has its critics.[4]

Unemployment is a serious problem in developing countries. Typical rates for urban unemployment range from 10 per cent to 15 per cent of the total urban labour force. The rates for young people between the ages of 15 and 24 are even higher and can be twice the urban average figures. In addition, particularly in the rural areas, there is a great deal of disguised unemployment or underemployment; this is where people are employed seasonally or for other reasons are unable to work on a full–time basis. This subject continues to attract the concern of the development economists. One model often quoted in the literature is that due to Lewis,[5] who assumed a two–sector economy: an industrial sector, and a subsistence–agriculture sector having excess labour. The Lewis model has been modified and extended by Todaro and others.[6]

The model is a tool frequently employed by the economist. A model is a simplified representation of a real–life system in which the model builder attempts to incorporate the most important features of the system he is simulating. If successful, the behaviour of the model will give some insight into the working of the real system and enable useful generalisations to be made. Whilst a model has some interest as a shorthand summary of the way in which a particular economic system has developed, its real value is in its predictive potential. The model should enable predictions to be made on the way in which the system on which it is based might develop in the future, and also to predict the probable progress of similar systems which are at an earlier stage of development. The sophistication of the digital computer has enabled models of startling complexity to be set up and solved. A whole new discipline, econometrics, has grown up to deal with this field.

One subject on which the development economists seem to agree is the importance of 'growth', and its relation to 'savings'. It seems generally agreed that in order to build up G.N.P. —that is, to achieve growth—it is necessary to save a fraction of the G.N.P. and reinvest it. The rate of growth will be proportional to the amount saved. The rate of growth will also depend on the return from the investment or, in other words, the ratio of the return to the amount of capital involved.

Savings in developing countries average around 13 per cent compared to 20 per cent in the developed countries. In order to relate savings to growth it is necessary to know the relation between capital and the annual return on it. For example, if the rate of saving is 15 per cent and capital to annual return is 3, then total annual percentage growth will be $15/3=5$ per cent. In addition to a lower savings rate, the developing countries have a higher rate of population increase than that for the developed countries. If in the last example the population rise was 2 per cent, the annual growth/capita would be reduced from 5 to 3 per cent. Some economists regard low rate of savings, or low rate of capital formation, as the major reason for slow development.

Internal savings can be supplemented by overseas aid. Some authors feel that overseas aid is essential for development: others representing both left and right wing points of view are opposed to it, but for different reasons.[7] Bauer argues that aid inhibits development within the country itself, whereas Hayter believes it to be a form of imperialism and hence unacceptable.[7]

Discussion

The factual background to underdevelopment is fairly well agreed—namely, that the developing countries are characterised by some or all of the following

(1) Low G.N.P./capita.
(2) High birth rate.
(3) High unemployment.
(4) Heavy dependence on the agricultural sector.
(5) Inequality of income.
(6) Low rate of capital formation.
(7) Poor infrastructure. Transport systems, water supplies, electrical supplies, telephones, etc.
(8) Poor services. Health and education, etc.

Opinions differ on where the development effort should be placed within a country, some suggesting that the major effort should be in the predominantly agricultural rural areas, others that the build up of industry should have highest priority. There is also disagreement over the manner in which development should occur. Economists such as Kaldor[8] assert that capital intensive solutions should be adopted as the most effective way of raising G.N.P. and hence the general development level,

whereas Schumacher[9] believes in concentrating on the problem
of increased employment by the creation of more low–cost work
places, this latter view representing the Appropriate Technology
approach to development.

There does not seem to be any consensus of opinion on policies
for development amongst the economists, some believing in the
virtues of centralised planning and others in the capitalist
system, and with further differences of opinion on actions within the two systems. That there should be these divergences of
opinion is probably not surprising; the developing world
is not a homogeneous mass, but contains great differences
in state of development, political systems and social group-
ings. What is good for one country may not be right for
another. Also it is difficult to separate the political from the
economic questions, and in some of the writings the two are
inextricably mixed.

There seems to be general agreement on the importance of the
rural, predominantly agricultural, sector and some authors recommend the placing of more emphasis on job creation in these
areas. Low capital investment is also identified as a major cause of
slow development. How then does Appropriate Technology fit
into the world of the development economists? Appropriate
Technology is not an overall economic plan but, in the words of
Dr Schumacher, a mechanism 'to go places, starting from where
you are'. It is particularly concerned with the dualism within
developing countries, which is manifested by the small islands of
technological development in the general sea of stagnation and
accompanied by a drift of population from the rural to the urban
areas, to join the already large numbers of unemployed.
Appropriate Technology is a methodology of development
which takes account of social benefits and costs in addition to
purely economic factors; it offers a package of techniques and can
be applied in all development situations. In particular, the emphasis on the creation of work places where the people now live
provides a solution to what is probably the major development
problem. The produce of these newly employed people will help
to build up the capital resources of the country, and hence
promote further development.

Future trends

The present world population of 3,860 million people (1973) is

increasing at an average rate of 2 per cent per year.* Projections
suggest that it will reach 7,000 million by the year 2000. The rate
of increase in the developed countries is much less than the
average figure, 0.8 per cent per year, whereas for the developing
countries the rate is 2.8 per cent per year. The high rate of
increase in developing countries results in a population structure
such as that for India shown in figure 2.11, which includes the
population structure of the U.S.A. for comparison. The high
proportion of children to adults in such a population is very
marked when compared with the structure in the U.S.A., for
example. As someone remarked, 'the rich get richer, the poor get
children'.

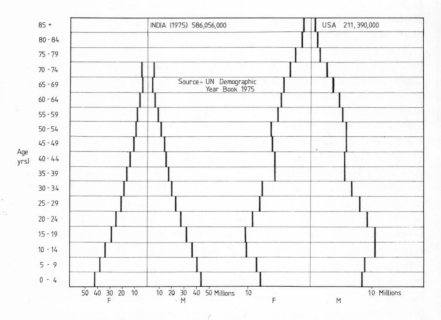

Fig. 2.11   Population structure—India compared with the U.S.A.

Doubling time

When populations grow, the rate of increase is usually given as a
percentage increase per year. This is difficult to visualise and a
more vivid way of expressing growth is in terms of the time

*If this increase were to continue, then, in rather less than 700 years, there would be one person for
every square foot of the earth's surface.

required for the population to double.★ This is called the 'doubling time' and may readily be obtained by dividing 70† by the percentage increase per year, the resulting doubling time being given in years. For example, if the population increase per annum is 2 per cent, the time required for the population to double is 35 years. The concept also applies to other quantities; for example, the current doubling time for the world's rate of use of energy is about 15 years. Hence, if this trend continues, in 60 years time, or four doubling times, we will be using 2×2×2×2=16 times more energy than we use today.

Figure 1.1 shows that the developed countries have around 20 times the income per capita of the developing countries, and this gap is growing.

Population and food supply

One consequence of population growth is an increased total demand for food. So far, over the last two decades, food supplies have increased, partly by increasing yield per acre and partly by extending the cultivated areas, with the result that food supplies have roughly kept pace with increase in population. The amount of cultivable land is finite and yields may be also expected to reach a limit. This suggests that eventually the world's population will grow to a level where it can no longer be fed, a limit first predicted by the nineteenth-century economist Thomas Malthus.

Energy and other resources

The world's energy use is increasing at about 5 per cent per annum, or a doubling time of 15 years. At present almost all our energy is supplied by fossil fuels (coal, oil, natural gas), the non-renewable energy sources. If present trends are maintained oil and natural gas may be expected to be exhausted in a relatively short period, perhaps 30 to 50 years, and coal after several hundred years. Fortunately, other, renewable, energy sources such as solar energy and virtually inexhaustible nuclear fusion are potentially available.

Other energy resources such as metals are also being used rapidly and are not renewable in the way that energy is, though in some cases substitutes are available. This increasing drain on finite resources poses serious problems for the future.

★ The rate of increase is assumed to remain constant over this time.
† Strictly, this should be 69·3.

Pollution and environment damage

Pollution and damage to the environment are side effects which arise from the increased use of energy and other resources. The result of this is a reduction in the quality of life and, in its more extreme forms, a threat to life itself.

The futurologists

Over the past decade there has been increased concern expressed over the possible results of continuous growth, and a number of studies have been made which attempt to predict future trends and identify possible dangers. One can distinguish two major schools of thought in this field, which, for convenience, I will describe as the Lemmings* and the Micawbers.†

The former group, the Lemmings, are believers in the continuous extrapolation of the exponential growth characteristic currently shown by such factors as population, resource use and pollution levels. The most notable exponents of this approach are King and his associates in the Club of Rome. Their publication, *Limits to Growth* by Meadows *et al.*,[10] describes a simplified model for world development with population, food, natural resources, industrial output and pollution as its principal components. The interactions between these components were included in the model and a series of predictions for the future were obtained, based on a range of growth assumptions. Six examples are given below.

(1)   Standard run. No change in current trends (figure 2.12). As industrial output increases, resource prices rise, less capital is available for investment, industry collapses, agriculture and health services decline. Death rate rises rapidly.

(2)   Unlimited resources. In case 1 resource depletion was the basic cause of the problem. In case 2 it is assumed that unlimited resources are made available using nuclear power to work the low grade ores. (Here ultimate disaster is caused by rise in pollution levels.)

(3)   Unlimited resources, control of pollution. In this

---

* Lemmings are small rodents which, from time to time, gather in great numbers and rush towards the sea; it frequently happens that in this headlong rush many fall from cliffs to their death. Animal behaviourists have theories to explain this strange behaviour. My own belief is that their leaders have an uncritical faith in continuous extrapolation. In fairness to the futurologist 'lemmings' one should note that their predictions usually are of disaster.
† Mr Wilkins Micawber, the incurable optimist in Dickens' *David Copperfield*, who never despaired but always believed that 'something would turn up'.

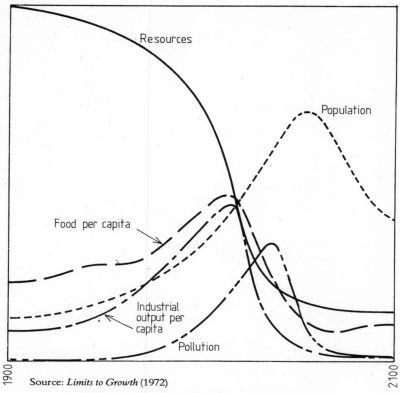

Resources

Population

Food per capita

Industrial output per capita

Pollution

1900

2100

Source: *Limits to Growth* (1972)

FIG. 2.12   World model: standard run *(Reference 10)*

model pollution per capita is assumed to be kept below
a fixed level. The limit to growth is found to be the
food/capita.

(4)   Unlimited resources, control of pollution, increased
agricultural production. In this case the limit is set by
pollution, even though pollution/capita is controlled;
the increased population causes a disastrous increase in
total pollution levels.

(5)   Unlimited resources, control of pollution, increased
agricultural production, population controls (there are
now only 'wanted' children). The limit is set by food
supply (figure 2.13).

(6)   Stable model (figure 2.14). A stable condition is
achieved with 100 per cent birth control, families
limited to 2 children, and industrial output/capita
restricted to the 1975 level.

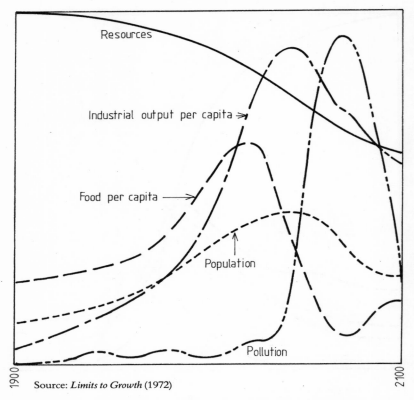

Source: *Limits to Growth* (1972)

FIG. 2.13   World model: unlimited resources; control of pollution; increased agricultural
production; population control. *(Reference 10)*

The Meadows model has been criticised by a number of
authors. The main criticisms are

    (1)   That the basic information and relationships used are
         inaccurate.

    (2)   That market forces and particularly technological de-
         velopments are ignored.

    (3)   That the model assumes the world to be homo-
         geneous.

Whilst it may well be true that some of the information
employed might be incorrect, the significance of the Meadows
predictions remains unchanged. The results with modified data
will still show the same overall characteristics as before.

The assumption of homogeneity in the model makes it difficult to identify choices and determine policies in particular

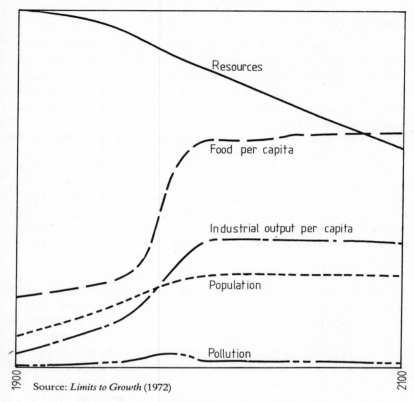

Source: *Limits to Growth* (1972)

FIG. 2.14   World model: stable model. *(Reference 10)*

areas. Mesarovic and Pestel[11] have set up an improved model in
which the world is divided into ten areas. They have produced
projections for these areas of a similar general nature to those by
Meadows.

The other major school of thought, the Micawbers, believe
that compensatory changes will occur and hence we will avoid
the disasters predicted by continuous extrapolation—that is,
they believe 'something will turn up'. Dr Kahn[12] of the Hudson
Institute is a leading figure associated with this approach, which
is sometimes described as the 'Technological Fix'.

The results of the various models do not predict what will
actually happen but only what will happen if certain conditions
are met. These models and predictions are of great interest and
importance in the planning of long-range development programmes.

The future

In order to develop, or progress, we must have an objective at which to aim. An improved quality of life for all is generally agreed to be the desirable goal for development. But what is meant by 'quality of life'? It is very difficult to try to quantify what are essentially value judgements — there is no international unit for quality of life.

For example, how should we compare the life style of the western executive with that of an elder in an African village? There will be large differences in the economic indicators between the two people. But what about factors such as position and esteem within the community? These are difficult questions to answer; what we can have no doubts about is the superiority of both life styles to that of the shanty-town dweller. Certainly there is a minimum level of resources and services below which life becomes unsatisfactory.

My old friend M. W. Thring has an interesting theory. He suggests that there is an optimum use of resources which maximises the quality of life; the graph in figure 2.15 expresses this in terms of energy use, but of course it applies to other resource use as well. He states that initially the quality of life will increase as the energy availability increases, in the sense that one will be kept warm, will have clothes, housing, food and other needs, all having an energy content. However, as the energy use increases, the quality of life will pass through a maximum and then decrease as effects such as pollution and ecological damage begin to become noticeable. Thring suggests that happiness is four tons of coal, per head, per year; whilst one might perhaps argue with him on the exact value, one sees the truth of his general thesis.

We have seen in the previous section that it is necessary to reduce the rate at which the world is at present using its resources, whereas in the developing countries it is clear that increased resource use is necessary to improve the quality of life. How are these apparently incompatible requirements to be met? A possible solution is for the developed countries to move towards a way of life which is more economical with resources without impairing, and hopefully improving, the quality of life. This will obviously be a difficult and slow process and the best that we can do in the short term is reduce the rate of increase in resource use, eventually reaching a zero growth which we should then attempt to reverse to reduce our absolute consumption (figure 2.16). In the case of the developing countries the

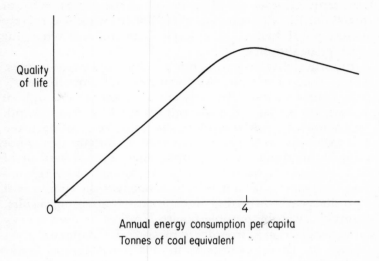

FIG. 2.15   Quality of life *v*. per capita energy consumption

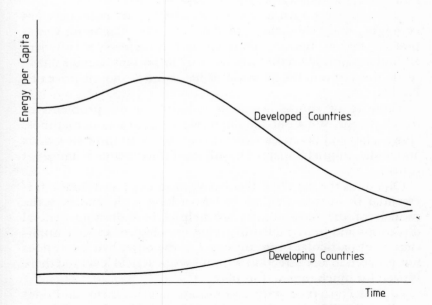

FIG. 2.16   Resource use goals for the developing and the developed world

problem is the reverse, that of stimulating growth over the whole population so that resource use will rise asymptopically to much the same value as that of the developed countries. What of the time scale? I have deliberately left this off since it is merely speculation but I believe this graph does indicate what our long-term goals should be.

It is important that, in setting out development programmes, we should use other criteria of value in addition to purely monetary considerations. An early and well known example of such a criterion is that originally put forward by Adam Smith and later adopted by Marx. Here the value of manufactured goods is expressed in terms of their labour content; this is not applicable to all things, for example, the value of land or of artistic achievement. More recently the subject of energy accounting has become fashionable; here one lists the energy cost of the product with surprising and sometimes alarming results. For example, even the humble potato requires its own energy content in fertiliser, etc., in order to produce it. Chapman[13] gives a number of these examples in his entertaining book. Schumacher[9] distinguishes between four categories—namely, that goods should be regarded as being either primary or secondary, and that these respectively should be divided into non-renewable and renewable, and manufacture and service.

Developments such as energy accounting are extremely encouraging since they show that we are now beginning to appreciate that economic factors are not in themselves sufficient and other considerations may be more important. Lovins nicely made this point in his question 'What is the economic price of a Dodo?'

There is of course no single solution to the problems of development. What one can do, however, is to set out methods, approaches and practices which have worked in some situations and which might be adapted to suit the different circumstances of others.

One finds the usual conflict between strategy and tactics. It is essential from time to time to take a long look, and here the models of the futurologists are helpful in indicating general directions and in particular in giving warnings of adverse implications of particular programmes. On the other hand, one does not proceed in any direction without work at field level and there is need for much more effort here.

A.T. is concerned with community development, and does not have a political content or view. Nevertheless its application in a particular situation may have profound socio-economic effects and hence political consequences. No action is politically neutral, not even installing a pump, since the effect may be either to strengthen or weaken the existing social system. Questions such as land tenure are crucial to development. My own view is that political questions should be the business of the indigenous population. Whatever the system adopted, at the level at which we are working the same jobs will require to be done.

The Appropriate Technology approach to development would seem to be broadly compatible with the views of the futurologists and at least some development economists; how can it be advanced? The rest of this book is an attempt to answer this question and give some practical suggestions.

The development worker should be aware of the place of his efforts in the long-term development context but he should not allow himself to be diverted into spending too much time on generalisations. One foot at least should be placed firmly on the soil (usually several inches deep in either mud or sand). There is no substitute for actual work in the field.

3

Intermediate Technology
and Appropriate Technology

What's in a name? That which we call a rose
By any other name would smell as sweet.
                    Romeo and Juliet, W. Shakespeare

The term Intermediate Technology first came into general use
following the writings of E. F. Schumacher, and particularly
after the setting up by him of the Intermediate Technology
Development Group in 1965. 'Intermediate Technology' is now
part of the literature of development. Intermediate Technology
as conceived by Schumacher was in the context of an economist,
that is, he saw I.T. as the stage between the subsistence £1 per
work-place economy and the developed, several thousand
pound per work-place economy. 'Intermediate' unfortunately
has connotations of the inferior or the second rate and also
implies that it is a stage to something more desirable. For these
reasons 'appropriate' may be a more suitable choice and in fact
the I.T.D.G. journal has the title 'Appropriate Technology'
'Technology' too could be criticised as implying 'engineering
machinery' whereas in fact the Appropriate Technologist is
engaged with total development which includes social and cultural factors, and in practice may be concerned with management, accountancy and marketing rather than engineering or
technology. 'Socially Appropriate Technology' is one attempt to
meet some of these objections. 'Low-Cost Technology' also
appears in the literature. In the developed countries the increasing concern over the undesirable results of high technology such
as pollution and the drain on resources has led to the setting up of
'alternative' societies who adopt 'soft' technologies. The aims of
such groups usually include an attempt to minimise resource use
and ecological damage. In this book we are concerned primarily
with the small-scale applications in developing countries,

40

in this field. 'Appropriate Technology' is now generally used, and understood, and is substantially equivalent to the older 'Intermediate Technology'. Appropriate Technology, or A.T., will be used as the preferred term in the remainder of this book.

Claims of Appropriate Technology

Appropriate Technology is concerned with all aspects of community development leading to a total or integrated development and hence to an improved quality of life for the individual members. Since most of the world's poor live in the rural areas, it is very much concerned with agriculture and agriculture-based activities. However, A.T. is not exclusively for the rural areas but is also applicable to the problems of the urban poor.

Appropriate Technology aims include

(1)   The provision of employment.
(2)   The production of goods for local markets.
(3)   The substitution of local goods for those previously imported and which are competitive in quality and cost.
(4)   The use of local resources of labour, materials and finance.
(5)   The provision of community services including health, water, sanitation, housing, roads and education.

It is important that such developments should be compatible with the wishes, culture, and tradition of a particular community and not have a socially disruptive effect.

Some objections to Appropriate Technology

In this section we will consider some criticisms which are frequently raised against the Appropriate Technology approach.
(1) The object is to build up a Western type technological and/or capitalist system. This aim is not inherent in the Appropriate Technology concept, though as in many other activities people who feel strongly about particular economic or political objectives will carry out their work with this in mind. However, what one should get clearly in perspective is that very many people in

the world today are living under conditions of abject poverty. These people have basic needs, such as access to pure drinking water (the consequences of the lack of this facility are very clearly brought out in table 5.1, p. 86). Poverty will not be sorted out overnight whatever the political or other changes, though one does not deny that in some areas of the world political change could be of considerable value. The solutions described in this book are applicable irrespective of political and other considerations, though their implementation could be very much helped by the right support from government. As far as Appropriate Technology as a concept is concerned, the actual development directions and objectives of a country are a matter for the people themselves to choose, and this choice will vary from country to country.

(2) Large-scale modern industrial methods are the only long-term effective methods to achieve development. This is not a question which can be answered by a simple Yes or No, 'It all depends on what you mean by . . .' as the late C. E. M. Joad was accustomed to preface all his remarks. The choice of a technology will depend very much on local conditions, resources and objectives. Modern airports, roads, hydro schemes and steelworks can under some circumstances be appropriate and in others not so. They are not necessarily appropriate because they are big, expensive, modern and prestigious. Detrimental side effects of large projects should be carefully considered, and as an example of this the Aswan project is discussed briefly in chapter 5. A good example of an inappropriate large-scale approach to rural development is the emphasis on rural electrification* in some of the developing countries which has not, in fact, benefited a significant fraction of the rural poor. In India 10 per cent of the electricity is used in the rural areas, where 80 per cent of the people live. Rural electrification efforts have been in progress in India for the past 30 years but only 30 per cent of the villages are electrified and even in these the utilisation is poor. For example, Makhijani[1] states that it is usual for only about half a dozen houses to be connected in a village of some hundred houses. He estimates that rural electrification has reached only about 5 per cent of the rural population in India and that these users tend to be the relatively well-off. The two levels of development are not

---

* I was given an interesting story of a different aspect of rural electrification by a campesino peasant in Colombia. He told me that he and his neighbours were no longer using electricity; apparently it had been the local practice to make illicit connections to the power distribution network. Unfortunately the power supply company had raised the voltage, making it dangerous to make the connection and in any case unsuitable for the domestic equipment.

necessarily incompatible but should be pursued in a complementary manner.

(3) Appropriate Technology is not new, only the name. There is of course a lot of truth in this statement. Innovative people have for many centuries made use of local resources to develop techniques and equipment for local conditions. A great deal of indigenous traditional technology is extremely well adapted to local needs,* and for many years people have been finding alternative solutions for the provision of modern materials and equipment. In his book *Doctor in Papua*, Vaughan[2] describes how he turned to the use of herbal remedies, making soap from coconut oil and wood ash, and surgical cotton wool from wild cotton when his normal source of supply was cut off at the outbreak of the Second World War. There are legions of similar examples in the literature. What is perhaps different about Appropriate Technology is that it is a total approach to development, that is, it covers all aspects of community needs and also, most important, it is concerned with continuous improvement in the quality of life and the development of the community. The international character of the recent interest in Appropriate Technology allows a sharing of information and experience which was previously lacking.

(4) Does it work? The answer to this is that it should, and it does sometimes but not always. This book gives a number of examples of failures since it is important to try to avoid these particular pitfalls in the future. Sometimes the failures are caused by the plant failing to achieve its specification, for example, the small capacity methane generators which were installed in an area where the average temperature was too low to give the required performance. More often the failure is not because the equipment does not give the required output but because the initial specification was at fault. The introduction of solar cookers to replace cow-dung cooking fires was unsuitable for several social reasons. Traditionally cooking was carried out in the evening; the women, understandably, objected to cooking in the midday sun, and finally the food did not taste the same. The peculiar flavour imparted by traditional cooking was absent. A more common adverse result is that of social disruption, the detrimental effect of a new development on other individuals or groups. This social cost is more difficult to predict and must be borne in mind when considering an apparently desirable change. An

---

* We are often surprised to discover that our 'new ideas' have been known and used for centuries, in some other part of the world.

example of the adverse effects of a change on a few individuals is given on p. 91 which describes the loss of employment of water carriers following the sinking of wells. The example quoted on p. 16 describes the widespread reduction in living standards of a community following the change from traditional leather shoe manufacturing to plastic moulding. The effect of the Green Revolution on whole areas in Pakistan is discussed in chapter 4.

On the other hand one can point to specific successes, spider glue and the Technology Consultancy Centre, Kumasi; Winner Engineering in Bangkok, Thailand; village polytechnics, Kenya; the Futuro method in Colombia; barefoot doctors in China, and paramedics in Peru and Colombia; Cretan mills in Ethiopia; leaf protein in Ife, Nigeria. There are many other examples of successes which could be given in India, Pakistan, Sri Lanka and other countries, and some are referred to elsewhere in this book. The above examples have been selected deliberately to bring out the diversity of the sources of information and the practitioners of the subject; some of the people involved, for example, at Kumasi and Ife, would no doubt describe themselves as Appropriate Technologists, but this would not necessarily be the case for Winner Engineering nor for the Chinese government. Nevertheless all these sources of information and experience are most valuable in initiating and guiding work elsewhere.

Having the ideas

So far we have introduced some of the problems and outlined the Appropriate Technology approach to their solution. A number of specific examples of Appropriate Technology equipment have been described. How are such solutions reached? Having the ideas and, most important, implementing them, is both the technical challenge and the excitement of working in Appropriate Technology.

In tackling a specific problem the first thing to get absolutely clear is what in fact *is* the real problem. So often one is presented with a request that is an implied solution rather than a statement of need. Having isolated and fully identified the nature of the problem, it is important to set out criteria against which to measure and compare the various solutions. We should also try to identify and list the local resources and skills available.

We are all potential inventors and have latent creative abilities but, sadly, conventional education tends to inhibit invention.

We are taught that there is a correct way to do things and are not encouraged to question established practice by suggesting alternatives. It is said that the conventional way is always the 'best' way, but we should realise that this assumption of an optimum is based on very special circumstances which do not necessarily exist everywhere. A bullock–drawn plough is infinitely superior to a tractor–drawn implement if the tractor does not work due to faulty maintenance or lack of spares. A hand pump is better than a wind–powered pump if the windmill has blown down owing to failure to replace the oil in the gearbox. The introduction of the tractor and the windmill in these circumstances would be serious design errors, showing a failure to appreciate the full requirements of the job.

The main impediment to invention is that of getting rid of the inhibitions which have built up during formal education and training, and the more advanced the training the more difficult this becomes. This is not an attack on professional training; it is obviously essential to have the discipline and knowledge of such courses in order to work properly at a professional level. It is instead a plea for us to be alive to new circumstances and to be sufficiently flexible to adapt to them. In particular we must resist the temptation to force conventional solutions onto unsuitable environments. There is a need for the lateral thinking of Edward de Bono. Excellent discussions of the processes of invention and creative design are given in references 3 and 4.

Sources of ideas and information

There are several fruitful sources of ideas for the solutions to problems; they include

(1)   Traditional practices, either modified and improved or transplanted to new areas.
(2)   Old techniques revived (Victorian engineering).
(3)   Modern industrial processes modified and reduced in scale.
(4)   New ideas exploiting advances in materials and techniques.
(5)   The 'Do it Yourself' industry.

Examples from all these categories can be found elsewhere in this book.

Where an advanced community is suddenly deprived of its

accustomed resources and equipment, considerable initiative and invention is shown in finding alternatives. Examples of such ideas include

(1) The substitutes for conventional materials (ersatz) developed in Germany during the Second World War to overcome shortages due to the blockade.
(2) Innovation in the Biafra region during the Nigerian civil war.
(3) The alternative societies, for example in California, resulting in publications such as the *Whole Earth Catalogue*.

The higher education sector provides a large potential source of scientific and technical know-how together with considerable goodwill and willingness to help in Appropriate Technology problems. Some years ago I.T.D.G. set up a pilot study to investigate this possibility. A unit, the University Liaison Unit, was set up at Reading University by Bob Congdon, who served as the liaison officer. He circulated a list of projects, compiled from requests for help received by I.T.D.G., to some fifty establishments.[5] The response from both staff and students was enthusiastic and a number of projects were started as part of normal academic course training. The intention was that the Unit should provide a liaison and coordinating service between the working groups and I.T.D.G. After a promising start the Unit had to be closed down due to lack of funds. It is, however, hoped to revive it again in the future. A list of the project titles is given in appendix I.

Panels and advisory groups formed by consultants drawn from the universities, industry and commerce are used by both V.I.T.A. and I.T.D.G. The latter has a number of specialist panels, including agriculture, water, power, and rural health. These panels serve a number of purposes,

(1) Provision of expert advice.
(2) The initiation of research and development where this is felt to be required.
(3) Provision of consultants for overseas visits.

Information storage and communication

The task of finding methods for the storage of information in a form which is readily understandable by the extension worker, and conveniently available to him, presents a serious difficulty to the various A.T. organisations. It is discussed more fully by McRobie.[6] Soon after it was set up, workers in I.T.D.G. realised that there was a lack of a convenient single source of information on available low-cost equipment. To meet this need a buyer's guide, or directory, called *Tools for Progress* was compiled. This contained details of equipment with a price maximum of £100 U.K., though most items were considerably cheaper. Glossaries were provided in French, Spanish and Arabic and over 7,000 copies of the guide were distributed overseas. Following the publication of *Tools for Progress*, I.T.D.G. Publications[7] have brought out a mass of printed material including specifications and drawings of equipment, handbooks, catalogues and annotated bibliographies (figure 3.1). Similar publications have been

FIG. 3.1   I.T.D.G. publications *(Photo I.T.D.G.)*

produced by other A.T. organisations, including the *Directory of Appropriate Technology*[8] by the Gandhian Institute of Studies in India, the *Village Technology Handbook*[9] by V.I.T.A. in the U.S.A., the *Handbook of Appropriate Technology*[10] by the Brace Research Institute in Canada, and the *Liklik Buk*[11] produced by the Melanesian Council of Churches. I.T.D.G. now publish a magazine, the *Journal of Appropriate Technology*, which is intended to serve as an international forum for news, discussion and the publication of developments of general interest.

Other methods of communication include field projects, consultancies and the setting up of Appropriate Technology centres in the developing countries themselves. Such centres already exist in India, Pakistan and Ghana and are discussed in chapter 8 and some addresses are given in appendix VIII. Unlike high technology with its highly developed communications network, Appropriate Technology disseminates very slowly. People twenty miles away from a successful rural project will be unlikely to have heard of it. Various means for dissemination are adopted including field workers; also the introduction of new ideas through the schools can be very effective.

Particular difficulties in communication at field worker/ peasant level arise and are discussed in the next section.

Communication—pictorial illiteracy

Every picture tells a story—but not the same story. Development is a two-way process and it is essential that those participating should be actively involved in, and sympathetic to, the aims of the programme. It is necessary in the development process for information to be transferred between the participants; this is a pretty obvious remark, but sometimes its importance justifies the statement of the obvious. For this purpose much use is made of posters and other visual aids. The problem of illiteracy and partial illiteracy is fairly well appreciated by those engaged in the presentation of development information. What is less well understood is the problem of pictorial illiteracy. This difficulty is vividly presented by Fuglesang[12] in his book *Applied Communication in Developing Countries*. Fuglesang draws attention to the conventions adopted in pictorial presentation, the language of pictures, and its rules which are familiar to the literate but may be quite unknown to the illiterate. For example, the convention that figures some distance away are shown smaller than those in the

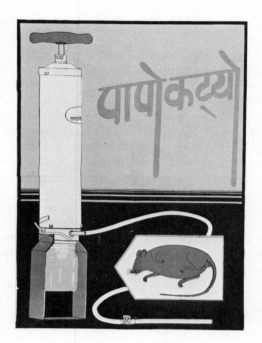

FIG. 3.2   Rat control: first poster *(Photo John Bowen)*

FIG. 3.3   Rat control: 'improved' poster *(Photo John Bowen)*

foreground can be interpreted, just as reasonably, as depicting smaller people. Perspective has formal rules which when understood are helpful but if not are merely misleading. Fuglesang points out that someone unfamiliar with pictures does not see the subject as a whole but is aware of individual parts, such as the tail of an animal or the feet, and may identify the animal by one feature alone. If a goat is drawn having a cow's tail, some members of a group will identify the animal as a cow. So it is important to get the details right. This gets even more difficult when the action of the animal is important to the message. Bowers *et al.*[13] give an example (figure 3.2) which shows a poster designed to initiate a campaign on poisoning rats that were destroying grain. 90 per cent of the peasant farmers questioned believed that the rats were asleep not dead, and hence saw no point in the poster. Figure 3.3 shows what was thought to be an improved version; this time the farmers recognised the live and the dead rats but concluded that the poster merely indicated that rats eat grain. In attempting to introduce new equipment designs one has a particularly hard task; the normal engineering drawing showing three projections—normally one front view, one side view and a plan—is quite unsuitable (figure 3.4). Three-dimensional sketches are more suitable but do require some skill in interpretation. Labelled photographs provide a simple, cheap and effective means of presenting information (figure 3.5).

Fuglesang once again raises doubts with his sketch of a hoe (figure 3.6). When a group of people well familiar with the hoe were questioned, 50 per cent failed to recognise the implement and some believed it to represent a man in a vee-necked shirt walking along a road. If you will look at this sketch again you will see that this is a not unreasonable interpretation, but not much help in a discussion on hoes. Models can help, but one more salutary story. A lecturer gave a talk on the tsetse fly and illustrated his remarks by reference to an eighteen-inch model of the fly. Following the lecture, one of those attending said that he quite understood that such flies could pose a problem, but the local variety were very much smaller, so the problems were different. Everyone involved in the business of lecturing will have met this trap—it is the fault of the lecturer not the lectured—and I have had the embarrassment of falling into it myself on a number of occasions. For example, after what I thought was quite a good but somewhat mathematical series of lectures, on the transmission of radio waves through hollow metal pipes, I discovered that although following the maths quite

First Angle Projection

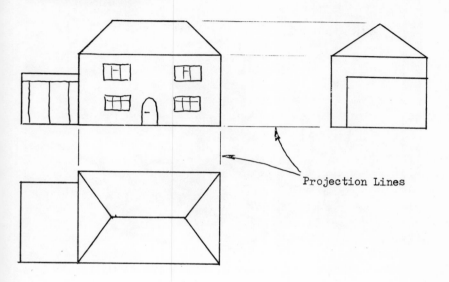

Projection Lines

Third Angle Projection

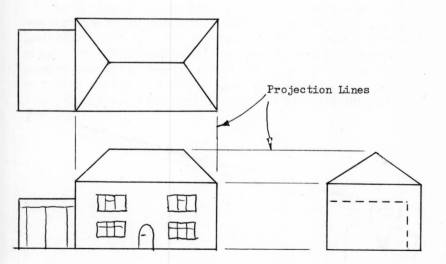

Projection Lines

FIG. 3.4   Engineering drawing

Land Drive Wheels
1¼"×¼"×24" dia.
Spokes
6 of ⁵⁄₁₆" dia.

Frame Cross-Member
of 1¼" Angle Welded
Box, 6¼" wide

1¼" Angle Box,
6¼" wide

FIG. 3.5   I.T.D.G. single-row rice seeds *(Photo I.T.D.G.)*

well the students believed the pipe to be a solid bar. Regardless of
my description, my blackboard sketch must have looked like a
solid bar.

The best method by which to communicate is undoubtedly by
example, by a field worker who understands the local people and
lives with them. Futuro (chapter 9) have tried the process of
recruiting local peasants on a part-time basis with encouraging
results. Nevertheless it is usually necessary to use printed
material and in preparing such information we must bear in
mind the pitfalls and try to avoid at least some of them.

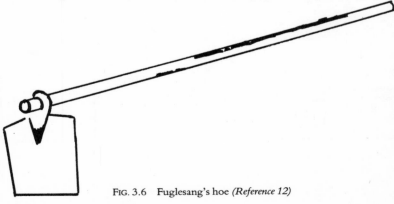

FIG. 3.6   Fuglesang's hoe *(Reference 12)*

Project selection

It has been stated several times that Appropriate Technology offers a complete development package for a community. This is so, in principle; however, in practice, because of limited resources and the relatively short period of operation, most work so far has been confined to single projects.

In selecting a project the full socio–economic implications must be borne in mind. First, the project must be of real benefit to the individual user. Secondly, the overall social benefit to the community must be positive; for example, a change must not result in unemployment elsewhere in the community. At government/community level, the benefits expected will include one at least of the following:

(1)   A positive contribution to the quality of life.
(2)   Import substitution.
(3)   Encouragement of new industry.
(4)   Building up of local skills.

These effects, in economists' terms, would be described as the micro and the macro respectively.

For example, let us consider the building of solar-heated crop dryers to replace either traditional drying in the open air or commercial dryers using imported oil. For the individual farmer it must be shown that the improvement in quality of the product and the avoidance of loss and spoilage on changing from a traditional practice is economically worthwhile, alternatively that the saving in fuel by changing from oil to solar energy more than offsets any difference in cost between the two equipments. The community benefits will include the setting up of local manufacture facilities and reduced oil import, and no negative effect would be likely to arise.

Scope of Appropriate Technology activities

From the earlier discussion A.T. has been understood to cover the provision of solutions to all the basic human needs. In particular it includes

(1)   Food, agriculture and agricultural engineering—new sources of food, improved agricultural techniques and the provision of equipment designs.

(2) Water and health—sources of water for drinking and other purposes, pumping, distribution and water treatment; also the safe disposal of sewage, and the provision of facilities which encourage improved standards of hygiene.

(3) Energy—though energy is not a specific human need, it is necessary to provide power for various human activities including water pumping, agricultural processes, crop drying, grinding, etc., transport, and for domestic and industrial use.

(4) Medical, building, roads and other services—the education and training of paramedical staff to provide some level of medical care to areas previously without any medical service; development of low-cost buildings and roads.

(5) Small industry—it is essential to build up small rural-based industries to provide the production and particularly maintenance of simple plant; this requires the encouragement of entrepreneurs by the provision of finance and advice on technical, marketing and accountancy skills, and similar help is required for the building up of the small urban industry.

(6) Education, training and development—education is an essential component of development and the systems followed in developed countries are not necessarily the most suitable or relevant to the developing countries. Traditional education has much to offer, and there are also some interesting new experiments in education. The training of more and better technicians is a general need in developing countries. The universities too can provide a lead in initiating and supporting local development. Research centres do not always work on solving the real problems nor in ensuring that their work is taken up by industry.

The above topics cover most of the Appropriate Technology interests and form the basis for the following six chapters.

4

Food, Agriculture and Agricultural Engineering

And he gave it for his opinion . . . that whoever could make
two ears of corn, or two blades of grass, to grow upon a spot
where only one grew before, would deserve better of mankind, and do more essential service to his country, than the
whole race of politicians put together.

Gulliver's Travels, Dean Swift

Most of the population of the developing world lives in the rural
areas and is engaged in agriculture or agriculture-related activities. Essentially, agriculture is concerned with the collection
of solar energy and its use in synthesising a few simple inorganic
materials—principally water, carbon dioxide from the air and
nitrogen—into organic vegetable materials by means of the
plant. This is the bioconversion process on which all life
depends.★ The resulting vegetable material may be used directly
as food or can be fed to animals which in turn may be used as
food. As the elementary biology books point out, animals are
entirely dependent on plants for their food. In addition to their
use as food, plants have long provided traditional building
material in the form of wood, rushes, etc., energy for heating
and cooking in the form of wood; also fibres, cotton, oils and
rubber.

Wood, bamboo and other plant materials provide the basic,
readily available, raw material for much of the Appropriate
Technology equipment described here. One should not under-
estimate wood as a structural material (material scientists classify
it as a composite) since it compares well in some respects with the
best modern materials and for some applications has mechanical

---

★ This statement is not entirely true because there is a type of bacteria which obtain their energy by
oxidisation of chemical elements. One member of this class has some importance to us since it is
responsible for the nitrogen fixation in the roots of the leguminous plants and is referred to in the next
section.

properties superior to them.* Wood still makes up a large fraction of the energy supply in the rural areas of the developing countries. With the new interest in alternative, renewable energy supplies bioconversion is receiving increasing attention for the collection and storage of solar energy on both a small and a large scale. Vegetation can also be used as a chemical feedstock, and to provide essential oils. This chapter is concerned with the production of food, and the role that Appropriate Technology might play in improving the food supply, whilst at the same time improving the standard of living of the small farmer, and increasing rather than decreasing the number of rural work places.

First of all we will summarise a few essential facts about plant and animal requirements and those facts relevant to human nutrition.

The plant

For our purpose the basic functions of a plant are shown in the block diagram of figure 4.1. Water and carbon dioxide from the air are combined by photosynthesis, using the energy from the sunlight, to form carbohydrates. These carbohydrates may be further converted, using nitrogen and other minerals, to proteins. Oxygen is released to the atmosphere as a by-product of the photosynthesis. The carbohydrates may be used as cellulose to form the structure of the plant, stored as starch or, as at night, the process reversed to produce water and carbon dioxide and release energy for use by the plant. The result of all this to the farmer is the production of carbohydrate and protein for use as described in the previous paragraph.

Although almost four-fifths of the atmosphere consists of nitrogen gas, plants are unable to absorb nitrogen in this form† and it must be provided as a chemical compound from the soil. The leguminous plants such as beans and peas are an exception to this restriction and although they cannot themselves make use of atmospheric nitrogen, they have associated with their roots a bacterium able to convert (or fix) the nitrogen gas to a compound which the plant can absorb. Presumably the bacteria also

---

* Some materials are stiffer than others, a property known as Youngs' Modulus; if we wish to make a beam which is both stiff and light-weight we will be interested in the specific stiffness — Youngs' Modulus divided by density. For wood and steel this works out to be almost exactly the same value. For more details on the mechanical properties of wood and other composites see Gordon's book.[1]
† There are some algae, the blue–green algae, with this capability. These are important in provision of nitrogen in wet rice cultivation.

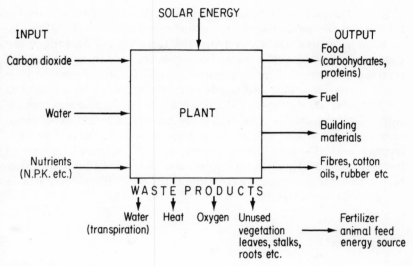

Notes:

    (i) Plants absorb oxygen at night and give off carbon dioxide but net effect is absorption of carbon dioxide

    (ii) Heat lost principally as latent heat of evaporation

    (iii) Transpiration accounts for typically 99% of the water initially absorbed by the plant

    (iv) Unused vegetation has a value as a soil conditioner, as a material whose breakdown products add to the plant nutrients in the soil, and, as a surface layer 'mulch' may assist in conserving moisture in tropical areas

Fig. 4.1   Block diagram of a plant

benefit in some way from their association with the plant, or they wouldn't do it, and this mutually convenient arrangement is known to biologists as symbiosis. The other minerals required by the plant include phosphorus, potassium, calcium, magnesium and sulphur together with smaller quantities of trace elements such as iron, zinc, boron, copper and manganese which are usually provided from the soil. Nitrogen, potassium and phosphorus are required in greater quantities than the other elements and must be added to the land regularly in the form of fertiliser to replace the material consumed by the plants. These elements are often referred to by their chemical symbols which are respectively N, K and P.

In traditional agriculture these minerals are continuously re-cycled by the return of human and vegetable waste. With the transport of produce away from the farm this is no longer the case and the minerals must be replaced, hence the need for fertiliser. Even the lush tropical jungle is only sustained by continuous recycling and needs fertiliser if it is to be worked as agricultural land. Since, in farming, a single species crop is normally required, it is necessary to protect the crop from enemies of various sorts; these include other plants (weeds), insects and disease. This protection will require the use of herbicides, insecticides and fungicides. The cost of these together with fertiliser constitutes a considerable financial burden particu-larly on the small farmer.

The animal

The block diagram for an animal is shown in figure 4.2. The input to the animal includes fuel and the oxygen with which to burn it (in a similar manner as for the car engine), material for structural and other purposes, and water. The fuel and structural material will be provided in the form of vegetation in the case of herbivores, and meat in the case of carnivores. The output of domestic animals will be either food (milk, meat, eggs), clothing (wool, leather, silk) or work in the form of muscle power. For humans, the useful output is in the form of work. It is interesting to look at the efficiency with which the fuel is converted to work

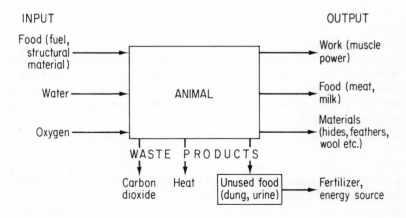

FIG. 4.2　Block diagram of an animal

in muscles. The gross efficiency is about 30 per cent, which is better than that of the car engine. The amount of muscular work that a human can carry out in a year (that is productive work, and does not include all his other activities) is around 200 horse power hours a year. If we divide this figure by his annual food intake in energy terms, we get an efficiency of about 12 per cent, which is not at all bad by comparison with conventional energy converters. In the next section we will look a little further at the question of overall conversion efficiency of the incident solar energy.

Food chains

Only about 1 to 5 per cent of the total solar energy falling on a plant is used in photosynthesis; the remainder is either reflected, or absorbed as heat and later lost to the surroundings. Of the carbohydrates produced, some become the 'useful' crop whilst the remainder are used by the plant in respiration and other internal processes, and in the production of 'useless' growth such as stems. The simplest way to use the crop is directly as food; alternatively the whole plant may be used as a feed for herbivores which in turn serve as a source of food. In this case, however, there are further losses since a substantial fraction, around 90 per cent of the energy of the food input is absorbed in respiration etc.★ Less efficient still is to use the cereal crop as cattle feed, as is common practice in the developed countries. For example, the average cereal consumption per head per year in the U.S.A. is 900 kg compared to a figure of 180 kg for India. This does not mean that the average American consumes fives times as much cereal as the average Indian, but rather that he consumes 70 kg of cereal as such and uses the remainder as feed for livestock, poultry, etc. These figures show the impact on cereal consumption of the meat component of diet in the developed countries. Currently the American diet includes 25 per cent of livestock compared with 17 per cent for Europe and 3 per cent for Asia.

When several levels (trophisms) are involved between the incident solar radiation and the final consumer, the sequence is called a food chain. Figure 4.3 shows a food chain having four levels. Referring to the figure it is seen that 50 per cent of the

★ Efficiency is not too meaningful as a criterion; for example, if cattle are fed on vegetable waste the feed is essentially free. A similar misunderstanding often arises in discussions on windpower, because efficiency of conversion is unimportant to the user and capital cost per unit of energy is the significant parameter.

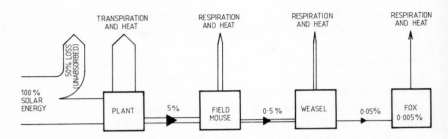

FIG. 4.3   Schematic food chain *(Reference 2)*

incident light is not absorbed by the plant. 5 per cent is assumed to be fixed by photosynthesis, and the remaining 45 per cent of the energy is lost mainly by transpiration. At the second level it is assumed that the plant is eaten by a field mouse which in turn fixes 10 per cent of the energy stored in the plant. If the field mouse is eaten by a weasel and this animal in turn consumed by a fox, the food chain is completed. Further 10 per cent conversion efficiencies are assumed in these two stages. The end result is that the carcase of the fox has stored only one twenty-thousandth or 0.005 per cent of the original solar energy. The conversion efficiencies given here are only approximate and will vary with different animals. However, the calculation is broadly true and brings out very clearly the undesirability of a long food chain. Figure 4.4a (from reference 2) shows a schematic five-stage food chain in which 300 trout are required to support one man for a period of one year. Tracing this back through the food chain of frogs and grasshoppers, we arrive at an annual input of 1,000 tons of grass. Figures 4.4a, b, c and d show, dramatically, how the numbers of people supported by this amount of grass will grow as the chain is shortened. In practice the food chains are far more complicated than those given in these simple examples, and 'food web' better describes the multiple interactions which occur.

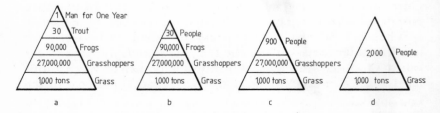

FIG. 4.4   Pyramid of numbers in a food chain *(Reference 2)*

Human nutritional requirements

Carbohydrates and fats provide the major part of human energy needs. Proteins, which are made up from different combinations of amino-acids drawn from some 20 naturally occurring types, are required to perform a number of important functions. Proteins are used for structural building, the manufacture of enzymes and nucleic acids. If they are provided in excess of requirements for these purposes, they are used as an energy source as an alternative to carbohydrates. The small quantities of minerals necessary are usually found to occur naturally in the diet, though this is not always the case where diets are of a restricted nature. Vitamins, though only needed in small quantities, are an essential component of diet. Finally around 1.8 to 3 litres of water a day is consumed; this is supplied partly by the food (table 4.1). There is some uncertainty in laying down

Table 4.1 Human nutritional factors

Requirement	Factor
large amounts	carbohydrates, proteins, water
small amounts	vitamins, minerals
minute amounts	trace elements

absolute minima for a diet since individual requirements will vary widely from one person to another according to weight, sex, age and nature of work. Also there are obvious differences in national requirements, for example, between the Asian peasant and the European. The former has accommodated to a diet which would be quite inadequate for a European. Whilst the diet of the Asian peasant in sufficient to sustain life on a continuous basis, it is not sufficient for the achievement of full capacity, either mentally or physically. The F.A.O./W.H.O. expert committee have laid down general guidelines on energy and protein requirements.[3]

Figure 2.2 showed the daily food intake for the developed and developing countries, broken down under calories and protein. As elsewhere, averages conceal large variations, and statistics have to be treated with caution. In particular, children, nursing mothers and the old may have a diet well below the average. It is not essential to include meat in the diet, as the proponents of vegetarianism are always telling us; it is necessary in that case, however, to eat at least two different kinds of vegetables in order to ensure a supply of the eight or so proteins which the body is unable to synthesise for itself from other protein.

Medical effects of dietary deficiency

It has already been pointed out that a significant fraction of the world's population subsists on a diet whose calorie and particularly protein content is below the minima recommended by the F.A.O./W.H.O. Whilst it is clear that such people are able to sustain life on this diet, its effects can be both harmful and significant. Children show retarded mental development and brain damage which is irreversible. Adults are smaller and apathetic, and people of all ages are less resistant to disease.

When the calorie and protein levels fall even further, clinical symptoms become apparent, particularly with the children. Undernutrition is the general term given to a shortage of calories (or energy content) in the diet, and may or may not be accompanied by a deficiency of other nutrients. Marasmus is an extreme condition of undernutrition leading to starvation and death. Malnutrition is the description applied when the body is not supplied with the correct amounts of certain nutrients. In developing countries malnutrition mostly refers to a deficiency of one or more nutrients. In developed countries cases often occur due to excess food. Kwashiorkor is a condition caused by lack of protein, and can arise even when the total calorie content is high. For example, in the case of children living on sugar plantations and able to obtain as much sugar as they wish, such children are sometimes termed 'sugar babies' (table 4.2).

Table 4.2 Malnutrition

Types	Effect	
excess	obesity	
deficiency	undernutrition	marasmus
imbalance	p.c.m.	kwashiorkor

However, nutrition experts believe that generally where the content of a diet is adequate to provide the energy need there is unlikely to be a protein deficit. Exceptions to this are in the case of a diet restricted to very starchy foods, as in the case of 'sugar babies' and also for young children who may not be able to consume sufficient bulk in order to absorb the required amount of protein from a low protein food. Where the calorie content is inadequate but the protein content sufficient, some of the latter may be used for energy needs resulting in a net protein deficiency. Because of this interaction between energy and protein needs, it is now customary to refer to protein calorie malnutrition (p.c.m.) or protein energy malnutrition.

The F.A.O. estimate that, in 1970, 460 million people suffered from severe protein energy malnutrition. A large number of the severely malnourished are children under 5 years of age.

Common vitamin deficiencies include that of vitamin A; this can cause Xerophthalmia (blindness) which occurs extensively in Africa and India. (F.A.O. estimate that 100,000 children in the Far East go blind each year from this cause.) Beri–beri is also of wide occurrence and is due to vitamin B1 deficiency. Its incidence has increased markedly following the introduction of rice polishing which removes the husk containing the vitamin. Figure 4.5 (from reference 4) compares the vitamin and mineral

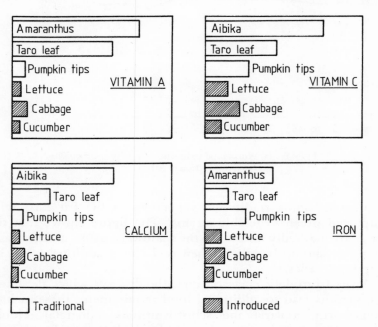

FIG. 4.5 Comparison of vitamins and minerals in traditional and introduced green vegetables in Papua, New Guinea *(Reference 4)*

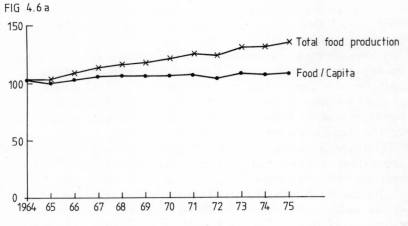

FIG. 4.6   (a) World food production *(Av. 1961–65=100)*
(b) Food production in Africa

content of some traditional vegetables in the Far East with imported green vegetable strains. The figure shows that the latter are markedly inferior to the traditional plants and indicates that care should be taken when replacing traditional foods by imported varieties.

Apart from deficiencies in diet, the diseases which debilitate the victims lead to additional food requirement; these include hookworm, malaria and schistosomiasis (bilharzia). Hookworm, for example, causes loss of blood which must be made up from the food intake.

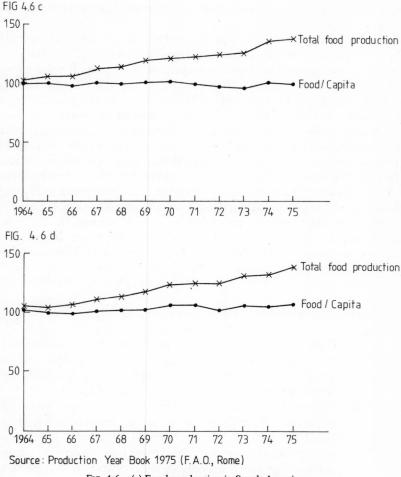

Source: Production Year Book 1975 (F.A.O., Rome)

FIG. 4.6   (c) Food production in South America

                  (d) Food production in Asia

World food supplies

Total world food production has increased by 32 per cent over the past decade (1965–75). However, this increase has been accompanied by a population increase of 24 per cent which has very largely neutralised the effect of the increase in production, and in some cases food supply per capita has actually fallen slightly. The resultant increase in food per capita over the whole world for the last decade is only about 8 per cent. Figures 4.6a, b, c and d show how the total food supply and food per capita has varied over this period for the world, Africa, South America and Asia.

We are led to conclude that agriculture is not achieving much more than the maintenance of present nutritional standards. Since the world population is expected to double by the early part of the next century, the world food production must also double in the same short period if even the present standards of nutrition are to be maintained. How is this to be achieved?

Whilst some new land will be brought under cultivation, most of this increased food supply will need to be produced on land at present under cultivation in the developing countries themselves, that is, from the areas where the people live. A great deal of the farming in developing countries is concerned with the basic food crops, cereal grains, root crops and food legumes. Unlike the plantation crops such as sugar and rubber, these crops have, until recently, received little development attention. Efforts are now taking place in various centres to improve the plant strain to give higher yield. Other developments in food production which may be expected to take place include

(1)   Improved use of land and water management.
(2)   The increased use of multiple cropping systems.
(3)   The nitrogen fixing legumes have been mentioned previously. Since nitrogen is a major component in fertiliser costs, the development of nitrogen fixing non-legumes is of considerable importance.
(4)   Increased protein availability. Plant genetics is being employed in order to increase the protein content of some of the basic food crops which have a relatively low protein content. The protein content of some cereals has been increased. For example, maize has been raised from 9–11 per cent to 10–12 per cent and the protein content of cassava has been raised from a value of less than 1 per cent to 6 per cent.

Fresh water fish farming, though hardly a new concept, is not yet widely used, yet it offers the prospect of a cheap and simple source of high protein food.

A high protein substance may be extracted from green leaves by a mechanical pressing operation. This process, which owes much to the pioneering work of Pirie,[5] can be carried out on a small scale with simple, cheap equipment. An example is given later in this chapter. The food chain illustrations given earlier indicate how important this source of protein could be.

Work is now being carried out on the conversion of hydrocarbon wastes to protein by microbial action. The microorganisms used include yeasts, moulds, bacteria and single cell algae. This method is usually described as the single cell protein process, S.C.P. The feed material includes sugar, molasses, vegetation wastes, crude oil and natural gas. The resulting high protein product is used as an animal feed. It is not yet clear whether the process can be carried out on a small scale.

Not all the methods referred to above will be suitable for adoption by the small farmer nor for production on a small scale. For example, some of the high-yield crop strains may have unacceptably high fertiliser and/or water requirements, this difficulty is discussed more fully in the following section on the Green Revolution.

We will now briefly summarise trends in modern agriculture. Agricultural development over the past few decades has been concerned with increasing crop yields, at the same time reducing man-hours per unit of crop. These objectives have been achieved by a combination of the following

(1)   The replacement of mixed farming by specialised activity.
(2)   The concentration on monocrops.
(3)   The employment of high fertiliser input.
(4)   Mechanisation.

This capital-intensive, energy-intensive farming is best carried out in large units and is undoubtedly successful in producing food at low cost. These practices are not necessarily the most suited for the developing areas, most of which lie in the tropical and sub-tropical regions. Further generalisation is unwise since the developing areas represent a very large range of states of development, and social and economic systems. Agricultural practices include the subsistence economies, intermediate economies, and the most modern. Much of the agriculture in Africa is by small owner-farmers, whilst in Latin America a great part of the land forms part of large estates and there is limited land access for most people. We are concerned in this book not with global solutions but with equipment and ideas which can benefit the small man with limited resources, though the result of their general application will be significant.

What happened to the Green Revolution?

The Green Revolution was hailed in the 'sixties as the answer to the world's food problems but this early euphoria has now waned and some authors have heavily criticised the programme.[6] The Green Revolution was the description of the programmes based on the use of high–yield wheat and rice developed respectively by the International Maize and Wheat Improvement Centre (C.I.M.M.Y.T.) in Mexico and the International Rice Research Institute (I.R.R.I.) in the Philippines. These varieties are characterised by a high yield (perhaps a factor of two greater than normal varieties under field conditions), they are less light-sensitive, and have a rapid growth enabling several crops to be grown in a year. To achieve these results a high fertiliser input is required, together with an adequate water supply. The crop must be protected by herbicides, fungicides and insecticides. The high fertiliser, herbicide and irrigation requirements make these crops capital-intensive though the yield to the grower results in an increased output which justifies the additional investment. The programme was started in 1943 in Mexico using the high-yield wheat, and proved to be very successful. It was later extended to India, Pakistan, Turkey and elsewhere. High–yield rice programmes were implemented in the Philippines, Taiwan, Sri Lanka and India. Some careful studies[7,8] have now been carried out on the results of these programmes. It has been shown, for example in Pakistan, that whilst the results for the larger farmers have been successful as predicted, the general social and economic effects in the area have been detrimental. The reduction in food prices from the larger farmers has resulted in the smaller farmer being unable to compete and many have gone out of business. The benefits of larger units have led owners to evict tenants to enable bigger farming units to be set up. The net effect has been increased rural unemployment. Nationally the drain on foreign currency caused by the imported fertiliser costs has also proved serious, particularly because of price increases due to rise in energy costs.

Whatever the final outcome of the controversy on the Green Revolution it is clearly not the total solution to the food programme neither is it an appropriate solution to the problems of the small farmer.

Mechanisation and its effect on work places

Mechanisation has real economic advantages for the large farmer; unfortunately it displaces labour, giving an overall negative social effect in many cases. For example, it is estimated that two and a half million jobs have been lost due to the introduction of tractors in Latin America. Another adverse aspect of mechanisation in a developing country is that it will almost certainly depend on imported equipment. This removes from the country what would otherwise be a significant industrial market for locally produced equipment.

Appropriate Technology equipment for agriculture

The object of the agricultural process is to achieve a high net crop yield★ for a given input of resources. The gross crop yield will be less than that theoretically possible, owing to various inefficiencies in production, and the net yield will be further reduced by losses which occur during processing and storage. Threshing, for example, can result in a considerable loss. Losses due to rats, infestation and spoilage are estimated to account for as much as a third of all African food crops. The extent of these losses during storage is not generally appreciated by the African farmers. Marilyn Carr[9] reports in a study in West Africa that the villagers were not really interested in the introduction of improved storage methods. On the other hand, the same villagers showed immediate interest in the introduction of a communal maize miller; previously the women were required to spend two or three hours pounding the grain in a mortar, so the advantages of the mill were immediately obvious to them. As we have stated several times elsewhere, the problem is one of communication, and is best solved by the training of local people who will actually work in the villages. There are no general solutions. Appropriate Technology is appropriate only to specific needs and cannot be transplanted without modification. It is necessary actually to live and work in a community in order to understand properly the needs and capabilities of the community and to gain acceptance for a change in practices.

★This will normally be in terms of cash return.

The small farmer is subject to a number of difficulties and restrictions such as

(1)   Labour bottlenecks.
(2)   Shortage of capital.
(3)   Land tenure. This problem often occurs. Why improve his land if as a result he may lose it?
(4)   Community and social pressures. Why work harder than your neighbours?
(5)   Land availability. He may not be able to expand due to lack of land. This is particularly a problem in the Far East.
(6)   Lack of knowledge of possibilities for improvement.

The problem of labour demand peaks is particularly serious. Figure 4.7 taken from Wijewardene[10] indicates the labour peaks

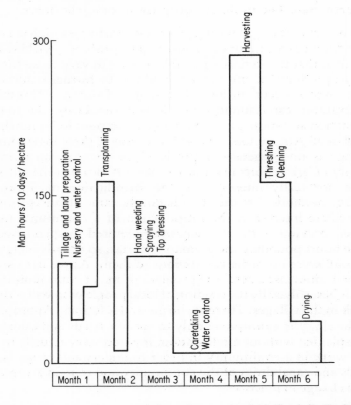

FIG. 4.7 Manpower peaks on typical one hectare rice field (flooded). Traditional farming. 120–130 day crop. (After Wijewardene.[10])

which occur in a particular crop in West Africa. In practice, these peaks tend to be somewhat smoothed out by activities connected with other crops, though peak labour requirements for two crops may coincide, and, for example, often the sowing of one crop is delayed from the optimum date due to the weeding load of another crop. In many cases weeding is the principal cause of labour peaks; Wijewardene estimates that 50–70 per cent of the total labour used in the humid tropics of West Africa is taken up by weeding. In a particular situation it is necessary to analyse carefully the whole farming process in order to identify present and future bottlenecks. An integrated approach to small farming practice is required. It is no use flattening out one peak if it merely reappears elsewhere in the cycle.

The aims in applying Appropriate Technology are

(1)   To identify the bottlenecks, inefficiencies, sources of loss.
(2)   To provide simple, cheap, appropriate solutions.
(3)   To disseminate information on these solutions and encourage their use by local farmers.
(4)   To arrange for a local source of supply and maintenance, preferably by a local manufacturer.

Agricultural equipment will usually be operated either manually or by means of animal power. Traditional equipment has been developed for this purpose over many years and may still be the best solution, though quite small modifications to the design or construction can often be well worthwhile. In some developing countries this equipment is manufactured in the country and is readily available. An attempt to make this equipment more generally known has been made by the production of the I.T.D.G. publication *Tools for Agriculture*[11] which lists a large number of manufacturers of low-cost equipment. However, this equipment is not available in all developing countries, and in the case of animal-drawn equipment even the relatively low-cost items listed in the catalogue may be beyond the resources of many small farmers; there is therefore a need for equipment to be made very locally, using local resources of skills, materials and capital. The provision of local sources of manufacture also ensures the availability of spares and maintenance. In the next pages we shall give examples of simple designs which have been evolved to meet specific needs, and which will indicate the types of skill required for their construction.

Local facilities for the production and maintenance of low–cost agricultural equipment

In most communities carpentry skills usually exist; however, for the construction of agricultural equipment it is also necessary to work metal (mild steel). At this level the technique involved can be summarised as 'Heat it, then beat it'. The iron is heated to red heat in a forge and then hammered to the desired shape on an anvil. Only simple bending and flattening operations will be required for the equipment with which we are concerned here. Of course the traditional blacksmith has developed manual skills, judgement and precision of working of a very high order; such expertise is not, however, necessary at this stage. A forge consists of two basic components: a hearth on which coke or charcoal is burnt, and in which the workpiece is placed, and a source of forced draught of air, usually provided by either bellows or a fan, to stimulate the fire. Figure 4.8 (from reference 12) shows a very simple type of traditional bellows which is constructed from a wooden box having an open top and divided into two sections by a wooden partition. A horizontal iron tube is fixed into each compartment as shown. The top of the box is covered by a flexible membrane, for example a goat's skin, and two sticks are attached to the skins to stand vertically. By holding these two sticks and moving them up and down vertically, a bellows action is achieved. The hearth is constructed from a small dip in the ground surrounded by stones, and a cone made from clay is positioned so as to direct the air from the bellows into the fire (figure 4.9). The anvil can be constructed from a

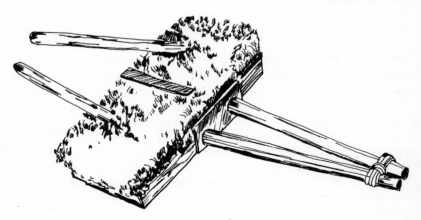

FIG. 4.8　Simple bellows *(Reference 12)*

FIG. 4.9 Simple forge *(Reference 12)*

sheet of iron nailed to a log. The I.T.D.G. group project in East Africa developed an anvil[13] using a short length of old railway line. If it can be obtained it serves the purpose most effectively (figure 4.10). A piece of rolled steel joist (r.s.j.) would be equally suitable.

A very neat design of forge, using a 45 gallon oil drum as the body, was developed by the I.T.D.G. Group Project in Zambia (described in reference 14). Two designs were constructed; the first employs a foot–operated bellows (figure 4.11a) and the second is designed to incorporate a simple fan (figure 4.11b).

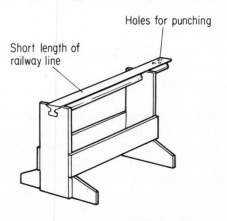

Holes for punching

Short length of railway line

FIG. 4.10 Rural anvil *(Reference 13)*

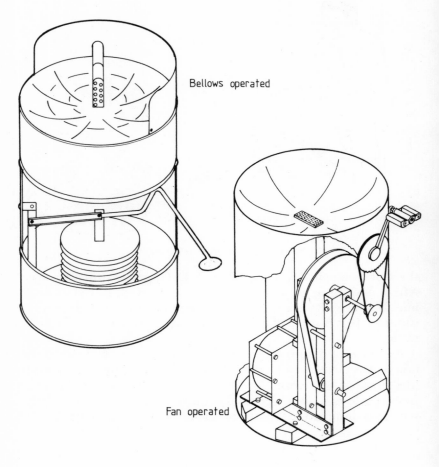

Bellows operated

Fan operated

FIG. 4.11   I.T.D.G. oil drum forges *(Reference 14)*

Agricultural processes

The conventional agricultural process is shown in flow diagram
form in figure 4.12. The first operation is that of land preparation. The land may require flattening, or if cultivation is to be
carried out on a hillside it may be necessary to arrange for
terracing to reduce loss of soil by erosion. The land is then
prepared for seeding by cultivation (ploughing and if necessary
subsequently breaking up the clods by harrowing). Once the
land is prepared, then seeds can be sown, followed in sequence

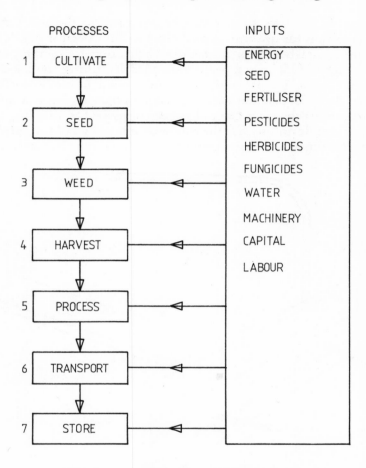

FIG. 4.12 Agricultural processes. Flow Diagram.

by weeding, harvesting, transport, processing and storage. The sequence of transport, processing and storage may not always be in this order. During the seeding/growing period fertilisers, herbicides and irrigation may also be required. It is important in a particular situation to examine each operation carefully before recommending a change since it is useless to eliminate one bottleneck if it is merely replaced by another at a different point in the cycle.

Land preparation

A scraper can be used for levelling out small humps of earth and filling in ditches and similar small earth–moving operations.

Figure 4.13 shows such a scraper designed by the V.I.T.A.[15] organisation for use with animal power. The operator stands on the horizontal board and the handle is held down for scraping and soil transport. The scraper is unloaded by lifting the handle. Where cultivation is carried out on sloping hillsides, surface water flow can result in serious erosion of the soil. For example, in Colombia the river Cauca has been described as 'bleeding the country to death', such is the extent of annual loss of soil carried away by this river. Soil erosion may be overcome by terracing,

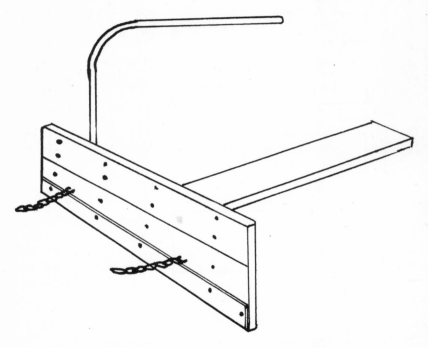

FIG. 4.13   V.I.T.A. buckboard scraper *(Reference 15)*

and is widely practised in many parts of the world. Figure 4.14 shows a form of terracing developed in the Medellin region of Colombia. This technique is interesting since it was initiated by a small farmer living in the area. This farmer was shown a photograph of stone–wall terracing in the Far East. This form of construction was not suitable for the Medellin area since sufficient stone was not readily available. The farmer devised a method of walling, using soil which he stabilised by the use of a network of roots of a coarse grass. This method has proved most

FIG. 4.14   Terracing in Colombia

successful in practice, and is now being adopted elsewhere in the region; farmers are invited to visit the terraces and are shown the method of construction, but it is not always possible to convince the farmers that the terracing justifies the considerable labour of its construction, even though they are surrounded by evidence of the loss of farmland by erosion.

Cultivation

Equipment for soil cultivation falls into three main groups; ploughs, cultivators and harrows.

Ploughs are fitted with a curved blade or mould board which both breaks up and raises the lower soil layer to the surface. With a fixed mould board the soil is turned to one side, usually the right. Since the plough is drawn backwards and forwards across the field the furrows will lie alternately from side to side. In the turnwrest plough it is possible to reverse the blade at the end of each furrow so that the soil is always turned to the same side. The ridge plough has two mould boards, making a 'vee' shape, which enables a ridge to be formed.

Cultivators have one or more chisel–shaped spikes which break up the ground but do not invert the soil.

Harrows are used to break up clods of earth and generally flatten the surface to prepare it for seeding. The tools vary and include discs, spikes and chains.

Figure 4.15 shows a number of ploughs and other cultivation equipment which are listed in the catalogue *Tools for Agriculture*.[11] In its simplest form an animal–drawn agricultural implement consists of a long bar, the tool bar, which is connected to the draught animals, and to which the tool blade, ploughshare or tine is itself fixed. The simplest implement will have a single tool; it is, however, usual to mount several blades on the tool so that a number of rows can be worked simultaneously.

Obviously the more tools, the more effort will be required to

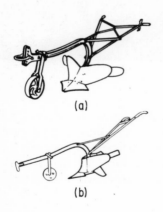
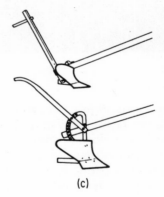

(a)

(b)

(c)

Key

(a) Ridger:
Ubongo Farm Implements
Dar–Es–Salaam,
Tanzania

(b) Emcot Ridger:
John Holt,
Zaria,
Nigeria

(c) Meston and Danish Mouldboard Ploughs
Danishman and Co.,
Lyallpur,
Pakistan

FIG. 4.15   Ox–drawn ploughs *(Photo I.T.D.G.)*

pull the implement. Provision must be made to adjust the height and angle of the tool, and where several tools are fitted the width must also be able to be varied. Reference 13 describes the Kabanyolo ox tool-frame (figure 4.16). This is an all-metal implement designed for manufacture and use in East Africa. The body can be used with a plough as in the figure, and can also be fitted with weeding tines or a seeder unit. There is a great deal to be said in favour of these multipurpose tool bars, particularly from the point of view of cost. Many farmers will have a plough and there is need for designs of tools which can be used with the plough. Reference 12 describes the construction of a simple all-wood harrow which is shown in figure 4.17.

FIG. 4.16   Kabanyolo ox tool-frame *(Photo I.T.D.G.)*

Seeding

Broadcast seeding is wasteful in the use of seed★ and makes subsequent weeding more difficult. Jab seeding is an improvement and figure 4.18 shows a simple implement developed in East Africa.[13] Further improvement can be achieved by the use of animal-drawn seeders having an attachment to measure seed flow.

★ This remark applies only to row crops.

FIG. 4.17   Wooden–toothed triangular harrow *(Photo I.T.D.G.)*

FIG. 4.18   Jab seeder in Ibadan, Nigeria *(Photo I.T.D.G.)*

Weeding

As has been stated previously, weeding is often the major labour load on a small farm. Where the crop is grown in rows an animal or man-powered weeder can considerably reduce this labour requirement. Figure 4.19 shows a mulching weeder.[17] These implements, which are ox-powered, are of simple construction and can be produced in a village workshop with only minimal facilities.

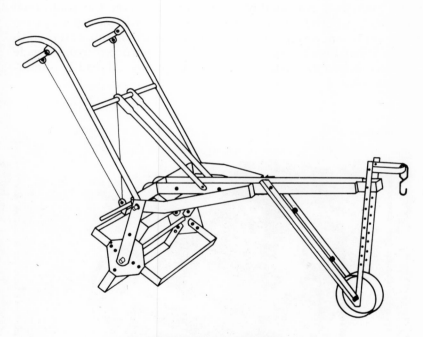

FIG. 4.19   Mulching weeder *(Reference 17)*

Spraying

Spraying of herbicides and pesticides can be carried out using back-pack equipment which costs less than one per cent of the cost of the tractor.

Irrigation

Equipment for irrigation is described in chapter 5.

Processing

Losses during processing can be significantly high. A traditional threshing method in Pakistan and elsewhere employs bullocks to pull a wooden sledge over the crop. The crop is laid in a circle on the ground and the bullocks walk around on it dragging a heavy wooden sledge. The grain falls to the bottom of the heap and about 60 per cent is recovered. However, an improved thresher is capable of processing about 100 kg/hr with a much improved recovery. Such a plant is relatively cheap, can be made locally and, whilst a considerable improvement over the traditional method, is both much cheaper and of more suitable capacity than a combine harvester.[18] Figure 4.20 shows a machine designed to strip kenaf fibres from the plant. It is constructed from bicycle parts and two discarded engine valve springs, all readily obtainable locally.

Fig. 4.20   Kenaf stripper *(Photo I.T.D.G.)*

Hot air drying using solar heating is a worthwhile improvement over the traditional method of laying out crops in the sun, since it enables the process to be controlled and avoids spoilage by birds etc. In addition, the high temperatures achieved (65°C to 70°C) destroy insects.

Some other possibilities

The above examples are taken from a wide area and are concerned with relatively simple modifications to conventional traditional farming practices. There are other possibilities which should not be overlooked. The variety and quality of the diet can be greatly improved by the adoption of kitchen-gardens and the introduction of rabbits* and poultry in areas where these are not familiar. Extension services are active in this field. Fish farming is becoming more generally popular as this source of protein has become more widely known. Less conventional solutions to the protein problem include the use of leaf protein, that is the extraction of protein from vegetation by crushing and extracting the juice. An interesting and encouraging example is reported by Oke,[19] from Ife, Nigeria, who says that green leaves processed mechanically by small-scale technology could be made to give a yield of 2,000 lbs of protein per acre (2,240 kg per hectare) compared to 700 lbs for soyabeans (785 kg per hectare) and 175 lbs for rice (195 kg per hectare). The leaves were pulped to disintegrate the tissues and then pressed to extract the juice. The juice was heated to 70–75°C to give a curd which can then be separated; alternatively it could be coagulated by slightly acidify-

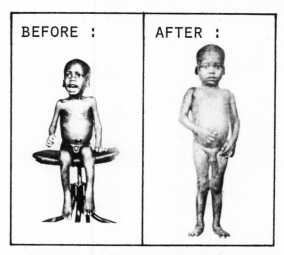

BEFORE : AFTER :

FIG. 4.21   Kwashiorkor patient *(Photo O. L. Oke)*

* Since some animals will inevitably escape, there may be a possibility of them adapting to the wild state and constituting a pest, as was the case in Australia.

ing (rather like adding rennet in order to curdle milk). The curd was then separated and dried. The taste of the resulting powder was 'rather like Chinese tea'. A large variety of leaves have been found suitable including leguminous plants (alfalfa, peas, clover, groundnuts, beans), cereals (wheat, rye, oats, barley, maize, millet, sorghum), grasses, vegetables, aquatic weeds, and other agricultural waste products (sugar cane tops, sugar beet tops, sisal waste, etc.). The resultant food product can be used either as a human or an animal feed. Experiments carried out in 1972 showed how kwashiorkor could be cured in a few months by adding this food to regular meals (figure 4.21).

Water and Health

There is plenty of rainfall in the world—in fact, if fairly divided amongst the world population we would all have around 28,400 litres per day* each.[1] However, rainfall is not uniform and does not necessarily fall where we want it to. Almost all communities live where there is access to fresh drinking water, and this usually means areas in which the rainfall is at least 250 to 400 mm per year (10 to 15 inches). The few exceptions to this are in desert areas where there are perennial rivers—for example, the Egyptian Nile. Where rainfall is seasonal there will be a storage requirement, also a transport need. For lifting from rivers and lakes, low lift pumps are used. If the water source is subterranean then bore holes and high–lift pumps will be needed. In the West we believe that water is something that comes out of a tap with unfailing regularity, but in the developing countries water supplies are a major concern and their provision is essential to the development process.

How much water do we need?

The daily water requirement of the average adult is between 1.8 and 3 litres per day. Of this, around 60 per cent is supplied from the food and 40 per cent from drinks. In order to maintain a balance, obviously the same average daily quantity must be lost. The corresponding breakdown is loss from the skin 15 per cent, respiration 20 per cent, urine 60 per cent, and faeces 5 per cent. The figure of 1.8 to 3 litres per day is of course the minimum amount to sustain bodily processes; in addition much larger quantities are used for domestic and washing purposes, for agriculture and in industry.

How much water do we use?

The U.K. domestic use is around 195 litres per day per head (the

* This is the total rain falling on the land surface which remains after evaporation losses have been deducted.

flush w.c. may account for as much as 37 per cent of this figure)
and in the U.S.A. the domestic use is around 300 litres per day
per head. In a developing country a minimum figure for domestic use is around 20 litres per head per day. The total water use per
day in the U.K. is approaching 300 litres per head per day and the
figure for the U.S.A. is several times larger, though much of this
water is returned to rivers relatively unchanged. Agriculture and
industry are major users.

Water and health

Water and health are closely connected; in fact the World Health
Organisation (W.H.O.) have estimated that 80 per cent of the
world's disease and illness is due to contaminated water. Water
can cause intestinal and parasite infection either by contamination, as in the case of drinking water supplies, or by providing an
environment in which disease carriers can flourish; examples of
the latter are the larvae of malarial mosquitoes and the snail host
to bilharzia. Lack of water results in a poor standard of personal
hygiene, which in turn leads to the transmission of infection by
means of unwashed hands, crockery, etc., and is responsible for
eye (trachoma) complaints and skin diseases. Water provides a
home for the larvae of disease carrying mosquitoes and flies, and
the extension of irrigation may well be accompanied by an
increase in diseases spread by these 'insect vectors'. Tinker[2]
quotes world totals for the number of cases of several diseases
and, in order to give a more vivid picture of the magnitude of the
problem, relates the numbers of victims to those countries
whose total population is of similar size (see table 5.1).

Table 5.1

Disease	Number of cases in the world (millions)	Country with total population equal to number of cases
gastro-enteritis	400	Non-Communist Europe
elephantiasis	250	U.S.S.R.
bilharzia	200	U.S.A.
malaria	160	Japan, Malaysia, Philippines
river blindness	30	Iran

Bradley[3] classifies the water–associated infective diseases under four categories:

(1) Infections spread through water supplies—water-borne diseases. For example, typhoid, cholera.
(2) Diseases due to lack of water for personal hygiene—water-washed diseases. For example, scabies, trachoma.
(3) Infections transmitted by aquatic invertebrate animals—water-based diseases. For example, schisto-somiasis, guinea worm.
(4) Infections spread by insects that depend on water—water-related insect vectors. For example, malaria, sleeping sickness.

To these must also be added a further group of infections associated with defective sanitation, for example, hookworm. The more common water–associated diseases and their sources are shown in figure 5.1. A more detailed account is given in appendix IV.

Water treatment

Much of the water–associated disease is related to polluted drinking water and, because of the high incidence of intestinal and parasitic diseases in tropical countries, a major source of water pollution is due to contamination by human waste.

The first line of defence is to avoid contamination by remov-ing the source, for example, by the siting of village wells away from latrines, and the drawing of river water upstream of the village sewage. A particularly likely source of infection is the use of buckets to draw well water, since these buckets are usually placed on the ground between immersions. This source of infection may be avoided by the installation of a simple hand pump. Hygiene education is another important component in the solution of the problem of water–borne disease.

The extent of the water treatment required will vary widely from site to site. If deep bores or springs are tapped where they appear from the hillside, no treatment will usually be necessary. Shallow wells and surface water are more vulnerable to pollution and will usually need treatment.

Even when the obvious sources of contamination have been

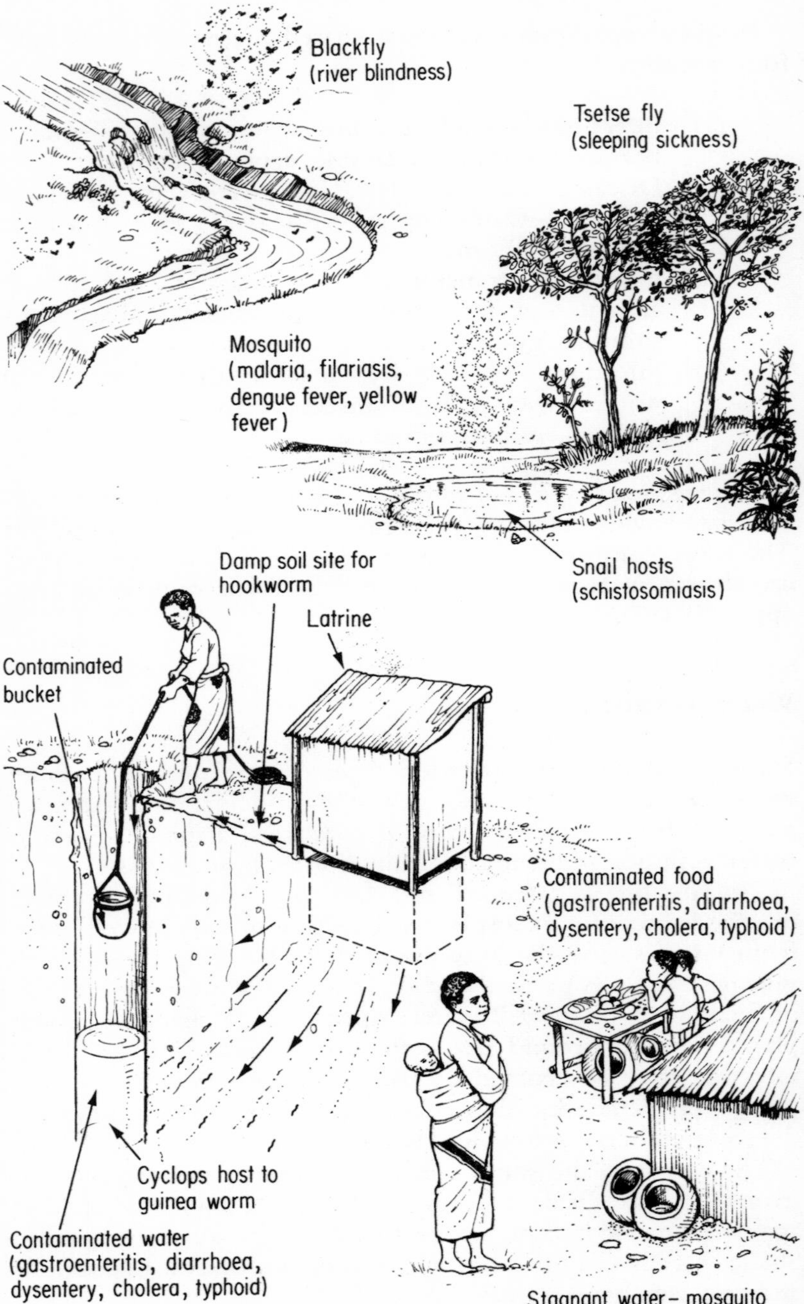

Blackfly
(river blindness)

Tsetse fly
(sleeping sickness)

Mosquito
(malaria, filariasis,
dengue fever, yellow
fever)

Snail hosts
(schistosomiasis)

Damp soil site for
hookworm

Latrine

Contaminated
bucket

Contaminated food
(gastroenteritis, diarrhoea,
dysentery, cholera, typhoid)

Cyclops host to
guinea worm

Contaminated water
(gastroenteritis, diarrhoea,
dysentery, cholera, typhoid)

Stagnant water – mosquito
(dengue fever, yellow fever)

FIG. 5.1   Water-related diseases

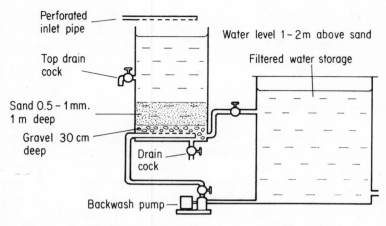

FIG. 5.2a   Rapid gravity sand filter[4] (2,400 to 7,200 litres per metre^2 per hour)

eliminated, it is often still necessary to provide water treatment
because once ground has been contaminated the pollution can
persist for years. Storage for 48 hours before use allows settle-
ment of microorganisms and the schistosomiasis larvae will not
survive this period. Sand filters can be used to take out suspended
solids. It is usual to distinguish two types of filter, the 'rapid' and
the 'slow'. Figure 5.2 shows versions of the two types. The fast
filter can be cleaned by back flow of the water; the slow filter
requires the sand to be replaced from time to time. Some
bacterial clean up occurs in the slow sand filter; it is desirable to
provide some disinfectant in addition to ensure that the water is
suitable for drinking. Chlorination is normally employed for this
purpose, either as a solution or in solid form as bleaching
powder. Simple methods for dispensing the correct amount of
chlorine can be set up.[4] Unfortunately, in practice, it is only too
easy to neglect this aspect by carelessness, or because of difficulty

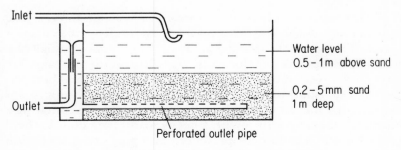

FIG. 5.2b   Slow sand filter[4] (100 litres per metre^2 per hour)

in supply, or cost of the chlorinating chemical. A very good description of water supply treatment is given in the I.T.D.G. booklet *Water for the Thousand Millions*.[5] Details of water analysis and treatment are given in reference 4.

Large-scale water systems

Since the Second World War many very big dams have been constructed in connection with either the generation of electricity from hydro power, or for irrigation or for both purposes. Examples include the Volta Dam in Ghana and the Aswan scheme in Egypt. These schemes are not unmixed blessings and may have detrimental ecological and other side effects; for example, in the Aswan Scheme the absence of the annual flood has seriously affected agriculture in the Nile Delta area and has virtually destroyed the Egyptian sardine industry which depended on nutrients entering the Mediterranean from the Nile. A serious health hazard has arisen due to the build up in population of the bilharzia–carrying snails in the sluggish water above the dam. Another difficulty is the ever present problem of silting. These are problems whose seriousness should not be minimised but which are not necessarily sufficient to offset the positive beneficial results of the scheme. Such developments may well be Appropriate Technology solutions and should fit in a complementary manner to the small–scale A.T. approach discussed in this book, since one outcome of the large dams is to make water supplies available to the small farmer who will then require simple lifting and handling equipment.

Rural water supplies

A typical water supply in a rural area may be a pond, river, lake or water hole situated up to several kilometres from the village it serves. Transport of the water in cans or other vessels, usually by women, represents a major daily work load. Other villages are served by a well or other supply in the village itself and here too there is a transport problem. (In some cases the water supply is at the foot of a hill on which the village is situated, necessitating a long and laborious climb with a heavy load.) Irrigation for field crops and kitchen gardens presents a pumping and distribution need. Cattle can usually walk to the source of water.

In 1970 of the 1,166 million population of some 90 'developing countries', 88 per cent, or 1,026 million, did not have access to a supply of pure drinking water.[1]

It is important not to overlook side effects—for example, the effects of a new development on other local groups. Colin Allsebrook told me of a borehole in an area in Ethiopia, where previously water had to be carried a considerable distance in skin containers on the backs of donkeys. This borehole was repeatedly vandalised by filling with rocks; this was thought to be the action of the men who had previously driven the donkeys and were now out of business. By appointing these men as well guardians the problem disappeared. This is a good example of the importance of taking into account sociological and other effects of an apparently simple technical improvement.

Irrigation

Water is essential for plant growth and very large quantities are required. 99 per cent of the water taken up by a plant is transpired and only 1 per cent is incorporated into the plant cells. One hectare of growing vegetation can transpire as much as 94,000 litres of water per day. In areas where there is a regular rainfall and which are situated in temperate climates, irrigation is not usually necessary. For example, only 3 per cent of the U.S.A. corn land is irrigated. This is not the case for most developing countries which are situated mainly in tropical climatic zones where rainfall is confined to a rainy season of three or four months and where, given a water supply, crops can be grown almost continuously. Water sources are frequently available but irrigation is not practised because of the capital and energy costs required. For example, only 10 per cent of the cultivated land in Africa is irrigated.[6] The energy cost of lifting the water will depend on its depth below the surface.

Equipment for supplying water is described in the next section. Having provided the water it must be used efficiently. Many crop varieties have been introduced to the developing countries from temperate areas where water supplies are readily available. Typically these plants consume 2,000 kg of water to produce 1 kg of dry matter. Work is being carried out on new, more water economical, strains. There are many techniques available for the more efficient application and conservation of water—for example, trickle irrigation, treatment of soils to reduce evaporation, more efficient cropping arrangements. These methods are discussed in references 7 and 8.

Water lifting and distribution

The principal sources of water are indicated in figure 5.3; to these we should also add the catchment and storage of rainfall. Wells and water holes are dug to reach underground water and have been known since the beginning of civilisation. The traditional

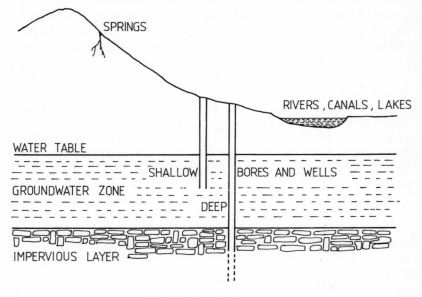

FIG. 5.3   Sources of water

well has a diameter of 1.5 m or more and is hand dug. In order to prevent the sides caving in it is usually lined with bricks or masonry. Seasonal, unlined, wells are frequently dug in river beds and elsewhere in the dry season in order to reach the water table. These temporary wells are unsafe and liable to collapse both in use and during construction. Wells have a number of advantages: they can be made using local unskilled labour, require only simple equipment, and may be constructed from locally available material. The use of the well is simple, requiring only a bucket on a rope, so there is no mechanical pump or engine to give trouble. Hand dug wells are not usually built to depths of greater than 75 m, though there are examples of wells of depths up to 150 m. Perhaps the greatest disadvantage of the hand dug well when used as a supply of drinking water is the danger of infection due to polluted water. The principal cause of this is the bucket; if the bucket is placed on the ground the bottom will pick up dirt containing various contaminants, including human faeces, which will in turn be transferred to the well. The problem is reduced if a concrete sill is provided and kept clean; it is also advantageous to extend the well wall for a couple of feet or so to reduce the possibility of contaminants being kicked or pushed into the well. Siting in the village is crucial—the well should not be near latrines—and an effort should be made during construction to render the upper walls impervious to water. The problem of disease is of great importance; it is not doing a village community a service to persuade them to give up the local stream, or whatever happens to be their water supply, in favour of a well in the village, if the result is merely to provide a source of infection which will ensure a fair distribution of all the local intestinal and parasitic diseases. I.T.D.G. have carried out work on the building of wells, using reinforced concrete, and have developed simple constructional techniques (figure 5.4).[9] The availability of cement and reinforcement rods for the I.T.D.G. design does not usually present any difficulty.

The alternative method of reaching underground water is by

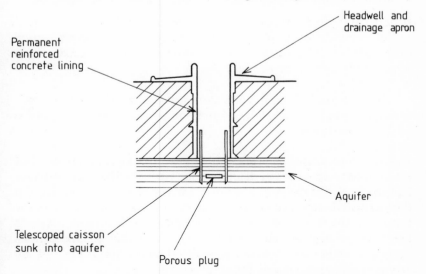

FIG. 5.4   I.T.D.G. hand dug well *(Reference 4)*

borehole (tube well). A borehole is a hole about 8 inches or so in diameter which may be sunk in one of several ways. Commercial rigs use power–driven boring drills, but a very simple rig can be used for manual boring. This consists of three poles tied together to form a tripod, a pulley, a hand–operated winch with a 'slipping rope' and a few rods, chisels and shells. (A shell is a steel tube with a flap at the bottom. It is used to remove loose material during boring.) In clay and loam, hand augers may be used. It is necessary to case the well to avoid collapse of the walls and steel pipe is generally used for this purpose. This is expensive and several local alternatives are possible. A simple casing can be made up using bamboo laths tied together with coir ropes; another possibility is to hollow out suitable bamboo type trees. More permanent casing can be constructed using short lengths of terracotta pipe; plastic pipe is now becoming available and its simplicity and long life often justify the higher cost over the traditional materials. One difficulty with all these materials is that the hole must be made and the casing lowered in. The casing cannot be driven like steel tubes. Boreholes need a pump since their diameter is far too small for a bucket to be used. This requirement removes one source of infection and thus gives the borehole a considerable advantage over the well.

Distribution

When the water supply is at a higher level than the required outlet, for example from a spring or a stream, water can be piped under gravity in plastic pipes or in gullies. In all other cases, some form of lifting will be needed.

(1)   Low Heads. Rivers, canals, lakes and shallow bores.
(2)   High Heads. Deep bores or where pumping uphill is required.

Figure 5.5 indicates the terminology we shall use. Total head or lift is taken to be the vertical distance between the water surface and the outlet of the water supply. This is made up of the sum of the vertical distance of the pump from the water surface, and the rise from the pump to the discharge point. The former depends on 'suction' and we shall see in a moment that the maximum vertical distance that water can be lifted by suction is limited, by atmospheric pressure, to about 10 metres.

In addition to providing the 'total head' pressure difference,

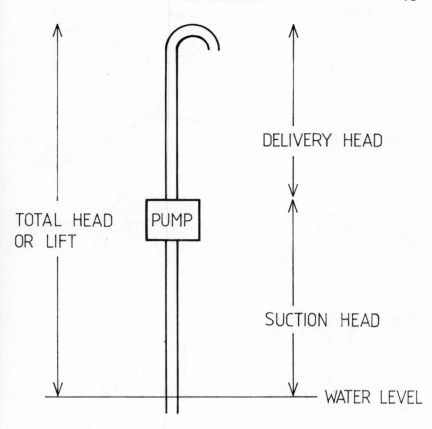

FIG. 5.5 Terminology

the pump will have to provide a pressure difference to overcome the friction between the water and the wall of the pipe through which it flows. For example, a flow of 0.5 litres per second through a 2 inch diameter polyethylene straight pipe is equivalent to 1.5 metres per 1,500 metre length. Additional pressure drops will occur at bends and changes in pipe cross section. Tables are given in the handbooks,[4,7] to enable friction head loss to be estimated.

Water lifting devices—pumps

Most pumps can be classified under one or other of the following three headings.

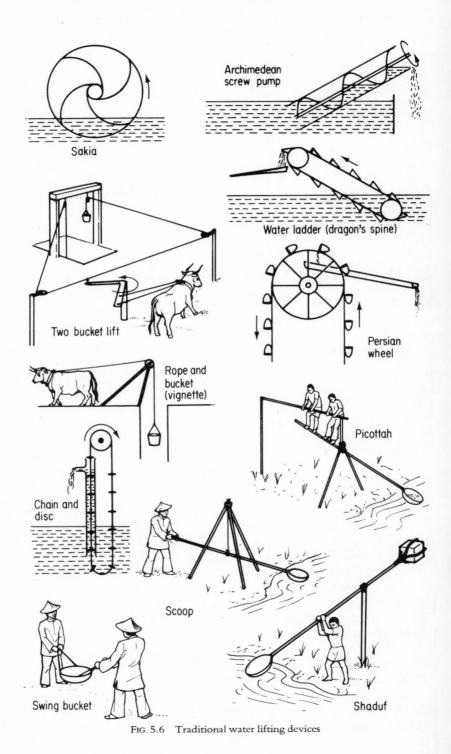

Sakia

Archimedean screw pump

Water ladder (dragon's spine)

Two bucket lift

Persian wheel

Rope and bucket (vignette)

Picottah

Chain and disc

Scoop

Swing bucket

Shaduf

FIG. 5.6 Traditional water lifting devices

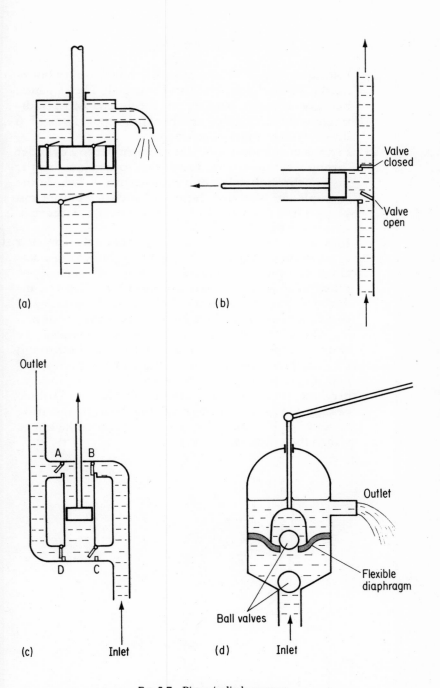

FIG. 5.7   Piston/cylinder pumps
(a) Hand pump   (b) Force pump
(c) Double acting pump (d) Diaphragm pump

(1)   Lifting devices—These are all different forms of bucket, which are filled, lifted and emptied. They include
buckets on ropes, buckets on wheels, etc. Most traditional water lifting devices are of this type.[11] Figure 5.6
illustrates a number of examples.

(2)   Pistons in cylinders—Water is induced into a cylinder
through a non-return valve and forced out by the
piston through a second non-return valve. Some
examples are shown in figure 5.7 and such pumps can
be operated manually or by other power sources such
as windmills.

(3)   Impellers—Most modern pumps are of the impeller
type. In these pumps a rotating blade pushes the water
along the pipe. Perhaps the commonest type is the
centrifugal pump shown in figure 5.8. Water is induced through a pipe on the axis of the pump and
caused to swirl round by the rotating vanes or impeller. The centrifugal force due to this rotation causes the
water to experience a force which drives it out through
the exit pipe. The detailed design of the pump will
depend on the head and flow which it is required to
produce; rotation at speed is a critical factor. Turbine
pumps consist of several centrifugal stages operating
in series, each stage adding to the potential height to
which the pump can deliver.

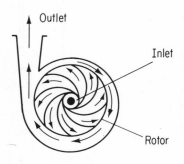

FIG. 5.8   Impeller/centrifugal pump

Suction

We have previously referred to suction and said that it is restricted to about 10 metres. We will now discuss why this
restriction occurs. Figure 5.9 shows a U-tube containing water

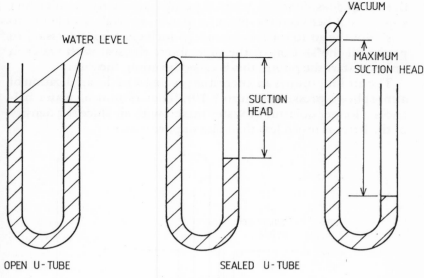

WATER LEVEL

VACUUM

MAXIMUM
SUCTION HEAD

SUCTION
HEAD

OPEN U-TUBE   SEALED U-TUBE

FIG. 5.9 Suction

and open at both ends. The atmosphere will press down equally
on each surface so that the height of the two columns will be the
same. If we now seal one of the tubes and pump out all the air the
atmospheric pressure will cause the water to rise in the closed
tube until it reaches the top. However, if the closed tube is made
very long there will come a time when the weight of the water
will exert a pressure equal to that of the atmosphere and the
length of water column in the closed tube will not increase any
further. The space above this column will be a vacuum. It turns
out that this height for water is about 10 m or 34 ft at normal
atmospheric pressure. It will vary with atmospheric pressure and
this variation is the principle on which a common form of
barometer operates. A 30 ft barometer is not a particularly
convenient instrument for the average household and it is more
usual to employ mercury as the working fluid.★ Mercury has a
density some 13.6 times that of water so the length of a mercury
barometer is only about 30 in. Returning to pumps, if we
remove the air from a tube then water will rise up to a height of
10 m; we say that it is 'sucked up' whereas in fact it is pushed up
by the atmosphere. It is necessary to exert a pressure equal to that
of the atmosphere to push the water out at the top, which is what

★ The vapour pressure of water at normal temperature is higher than that of mercury. For example, at
20°C, water is 0.02 bar but mercury only 0.0000013 bar, so that a water barometer would require
temperature correction.

the pump does. Since in practical pumps it is often difficult to get a good seal between the piston and cylinder wall, and hence to push out the air to start the pump, in such cases it is necessary to partly fill the section of the pipe above the seal with water in order to start the pump; this is called 'priming the pump'.

The air lift pump is an interesting example of the application of atmospheric pressure (figure 5.10). The column of water balances a longer column of water mixed with air since the density of the latter is much less than that of pure water.

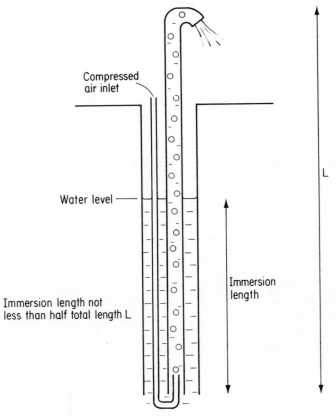

FIG. 5.10   Air lift pump

The I.R.R.I. bellows pump

This is a simple pump using canvas bellows which was developed by the International Rice Research Institute in the Philippines.[9] The pump is double-acting and is operated by the feet. It is used for irrigation and can lift 50 to 60 gallons a minute to a height of one or two metres. The operator shifts his weight from one foot to the other and alternately compresses the two bellows. Each is fitted with two single non–return valves as shown. One useful feature of the pump is that it can handle muddy water and small stones without incurring damage. The unit is equipped with a handle for carrying, and weighs only 20 kg. Figure 5.11 shows a similar double–acting pump with solid cylinders developed in China.[10]

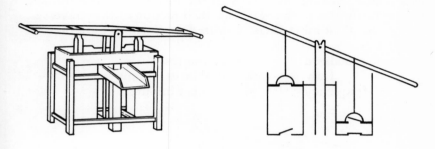

FIG. 5.11   Chinese double acting pump *(Reference 10)*

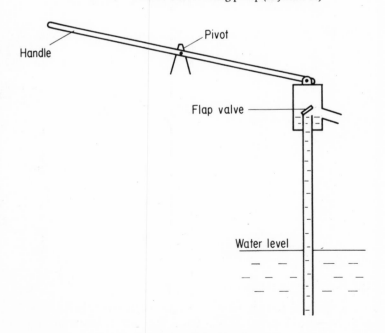

FIG. 5.12   Flap valve pump

The flap valve pump

This is a simple and popular design. Figure 5.12 shows its details. Such a pump can be constructed from locally available materials and none of the dimensions are critical.[9]

The resonant Afghan joggle pump

Figure 5.13 shows an improvement on the flap valve pump which has been in use in Afghanistan for over 200 years. A volume of air is trapped in the top of the pump cylinder and acts as a spring, storing the energy of the rising water column and then returning it to the water as the direction of flow reverses. Zapp, in Colombia, has reported performance figures as follows:[11]

suction head	1.5 to 6 metres
flow	60 to 100 litres per minute
frequency	80 cycles per minute.

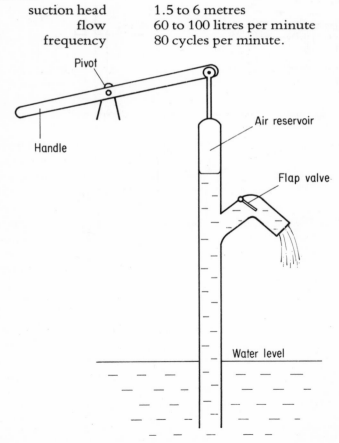

FIG. 5.13   Afghan joggle pump

The frequency was deliberately chosen to be about that of the human heart. Professor Zapp was surprised to find that very similar pumps, independently discovered, were already in use by the Choco miners in Colombia — nothing is new.

Power

In order to operate a pump a source of power is required. Power sources are discussed in more detail in chapter 6. It is instructive to see how much water we can pump with one horse-power. If the pumping process is fully efficient it is possible to pump 27,000 litres of water an hour up to a head of 10 metres (appendix V). Other situations can readily be derived from this figure; for example, if we need to pump to three times the head, the quantity pumped will be reduced by one-third. We see that even a man operating at, say, 1/20 h.p. for eight hours a day will pump a considerable quantity of water. In practice the pump will not be 100 per cent efficient and this figure may need to be reduced by a factor of as much as four. Even so, this is still an appreciable quantity. Animals have traditionally been used as power sources for water pumping. Wind powered pumps also have a long history. The Cretan mill design[12] has much to commend it (see chapter 1).

The I.T.D.G. windmill described in chapter 6 has been designed for pumping applications, and is intended for long life and simple maintenance. Modern methods of water lifting make use of either the internal combustion engine-driven pump or, where electricity is available, the electric motor-driven pump. Internal combustion engines can be of several types: diesel, petrol, gas or kerosene. The spark ignition,* internal combustion engines may be operated on bio-gas; the compression ignition engines (diesels) can also use this fuel but require some (10–20 per cent) diesel fuel to be used with the bio-gas. The internal combustion engines have the disadvantages of high initial cost, difficulties of arranging for adequate maintenance and of obtaining spares. Fuel cost is also a significant factor. The electric motor-driven pumps require less servicing but as in the case of the internal combustion engines both initial cost and fuel (electricity) cost is high. A serious disadvantage of both these modern solutions in most countries is that they are imported and hence use foreign currency; neither do they provide a market for local industry.

* In the spark ignition engine (for example, the car engine) the fuel/air mixture is ignited by a spark. In the compression ignition engine (diesel) the air in the cylinder is compressed by a volume ratio of around 16:1 and this compression heats the air. Diesel fuel is then injected and is ignited by the high temperature air. Bio-gas requires a higher temperature than diesel fuel for ignition and hence it is necessary to use a small proportion, 10–20 per cent, of the former in a compression ignition engine.

The hydraulic ram

If, when water is flowing along a pipe, a tap at the outlet end is turned off suddenly, then the momentum of the stopped column of water will cause a sudden pressure rise in the pipe. This effect occurs in domestic systems and is known as water hammer. The pressure rise due to water hammer is used in the hydraulic ram to pump water. The hydraulic ram requires a flow of water from a source and makes use of the flow energy of the water to pump a small fraction of it to a greater height. Figure 5.14 shows a schematic diagram of a hydraulic ram. The device is

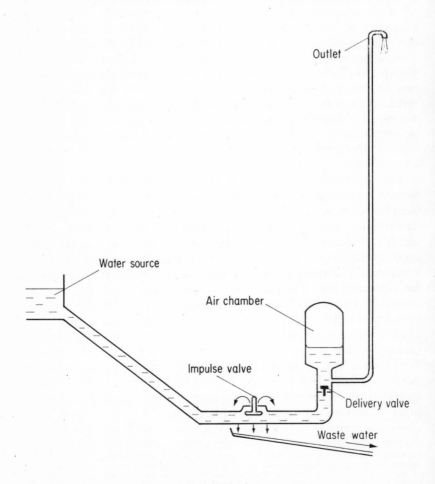

FIG. 5.14 Hydraulic ram

useful for pumping water from a stream up the hillside, and has the considerable advantage of not requiring a separate power source. Constructional details are given in references 7 and 13.

The Humphrey pump

The Humphrey pump[14,15] is a complete pumping engine combining both power source and pump. In an internal combustion engine, fuel and air is induced into the cylinder and when this mixture is ignited the gas pressure is increased considerably, and this pressure rise forces down the piston. In the conventional engines the piston is connected to a crank shaft by a connecting rod and the downward movement of the piston is converted to rotation of the crank shaft. This rotating shaft may in turn be connected either to the wheels of a car, or be used to drive a pump. In the latter case the pump also consists of a crank shaft which drives a piston in a cylinder, but here the cylinder contains water which is forced out by the piston to give the pumping action. Now this all seems a rather round-about method, and one might ask if it is not possible for the expanding gases in the first cylinder to act directly on a column of water—in other words, can we use the water as a piston? In fact, of course we can. Such a pump was invented by Humphrey at the beginning of the century. Humphrey built several very large pumps of this type with an output of 40 million gallons per day. Since 1969 at Reading we have been developing a smaller version of the Humphrey pump for use in developing countries. The operation of the pump is shown in figures 5.15, 5.16, 5.17.

Several pumps have been made; the performance of the latest, a six inch diameter cylinder pump, is as follows:

head	6.5 m
horizontal pipe	25 m
output	6 l/s
fuel	methane
efficiency (net)	10 per cent

Designs for these pumps have been provided for a number of organisations overseas, and I.T.D.G. has recently sent out two complete pumps, one to Nepal and one to Egypt, for field testing. The Humphrey pump is extremely simple to make; the piston does not require machining, does not wear, requires no lubrication. The pump can readily be made, operated and repaired locally.

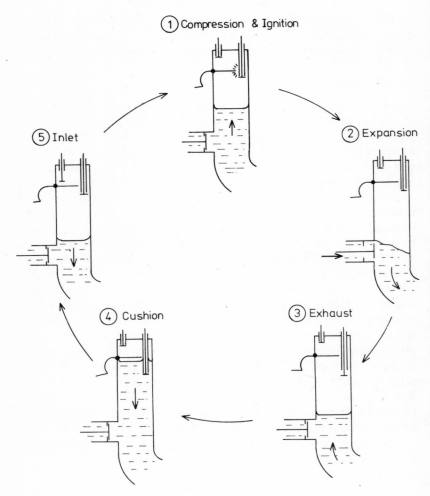

FIG. 5.15   Operation of 4-stroke Humphrey pump

One rather interesting use of the pump principle was the subject of an undergraduate project and concerned the use of the pump to drive a boat by directing the output water jet horizontally.

The Humphrey pump provides a good example of a research project suitable for a university in a developing country. The theoretical and instrumentation and design content is considerable and provides a well balanced Ph.D. type topic. The resulting source of expertise in the university is then available to local industry.

FIG. 5.16   Humphrey pump

FIG. 5.17   Humphrey pump installation

References 16 and 17 give two recent developments in pumps; one of these, the hydrostatic pump,[15] is a new idea, the second[16] is a revival of an old idea but with additional novel features. Both these pumps make good student undergraduate projects. Another interesting development concerns the pumps employing solar-powered vapour engines and reported by the Sofretes group.[18]

Collection and storage

Means of water storage are required either to store rain water or where the source is intermittent. An example of the latter is when windpowered pumps are employed. Small tanks (up to 200 gallons capacity) are constructed from concrete or resin-bonded fibre glass. Larger tanks can be built relatively cheaply using plastic or butyl rubber sheets. One such design has been described in chapter 1. Currently[1] I.T.D.G. are collaborating with the Ministry of Overseas Aid and Alcan Jamaica on the construction of a half-million-gallon capacity reservoir using butyl rubber sheeting.

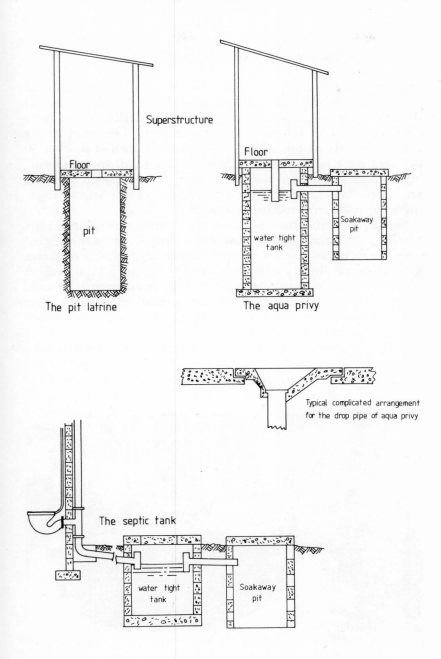

Superstructure

Floor

pit

The pit latrine

Floor

water tight tank

Soakaway pit

The aqua privy

Typical complicated arrangement for the drop pipe of aqua privy

The septic tank

water tight tank

Soakaway pit

FIG. 5.18   Pit latrine *(I.T.D.G.)*

Sewage, sanitation and waste disposal

Human waste is a serious source of infection and its safe and hygienic disposal is of great importance. More than 90 per cent of the rural population of the developing countries do not have adequate facilities for the disposal of excreta. Less than 30 per cent of the urban population have piped sewage and another 30 per cent have no sanitary facilities of any kind. In the towns, where high population densities occur, proper sewage facilities are essential for good health. In the rural areas self–help arrangements provide an immediate solution.

Simple constructional techniques for latrines are given in the handbooks[4,7] and figure 5.18 shows examples of such designs. The health hazards resulting from siting latrines near to water supplies have already been mentioned. Figure 5.19 shows an extreme example of bad site selection. The installation of ceramic ware enables a water seal to be introduced, which is a great improvement. The design and production of suitable ceramics is described in references 7 and 19.

Simple techniques for waste disposal are given in reference 4.

Fig. 5.19   Dangerous siting of toilets *(I.T.D.G.)*

6

Energy

The current total world energy use is around 6,800 million tons of coal equivalent. Our world energy budget is of course made up from several different energy forms, the most important of which are coal, oil and natural gas. It is however convenient to express them all in terms of the equivalent amount of coal. This unit, the tonne* of coal equivalent, or t.c.e., is easy to visualise, we all know what a tonne of coal looks like, and it turns out to be about the right size to express personal annual energy consumption. The world use of commercial energy per year per capita is 1.9 t.c.e. The rate at which energy is used is called power† and the units we shall use are the kilowatt (kW) and the horse power (h.p. approximately $\frac{3}{4}$ kW). T.c.e. per year is numerically very nearly the same as kW, so our average world energy use of 1.9 t.c.e. per year is the same as a two–bar electric fire burning continuously. In these units the energy equivalent of our food is 0.17 t.c.e. per year or about 1 lb of coal a day; in power terms we are all equivalent to two or three 50 W electric light bulbs burning continuously. Most of this energy appears as heat.

The world average figure of 1.9 t.c.e. per capita conceals very large national differences in energy consumption. For the U.S.A. the figure is 12 t.c.e. per capita, Europe around 5 t.c.e. per capita, whereas in the developing countries the figure is nearer 0.5 t.c.e. per capita. Now what has all this to do with development? It is found that there is a close relationship between commercial energy use and G.N.P. (figure 6.1). This is not surprising since a technological society requires large amounts of manufactured goods, all of which need energy for their production. Figure 6.1a (1969) shows that this relationship is roughly linear, with the U.S.A. at the top, Europe in the middle, and the developing countries clustered around the zero level. The U.S.A. with only 6 per cent of the world population

* The tonne or metric tonne is equal to 2200 lb or 1000 kg.
† For those who have forgotten their school physics the basic facts about energy, power, etc., are summarised in appendix V.

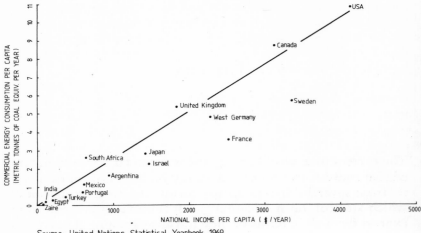

Source: United Nations Statistical Yearbook 1969

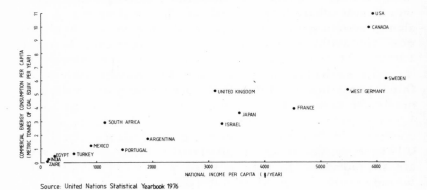

Source: United Nations Statistical Yearbook 1976

FIG. 6.1   National income per capita $v.$ commercial energy consumption per capita
(a) 1969 (b) 1974

consumes 35 per cent of the world energy demand. This is not a criticism of the U.S.A. but simply an indication of the energy intensiveness of a modern developed economy.

Figure 6.1b which refers to 1974 shows broadly the same pattern but there is evidence that some developed countries are now finding it possible to increase G.N.P. per capita without a corresponding increase in energy per capita.

At first human energy needs were provided by muscle power, both human and animal, and these power sources are still of prime importance in much of the developing world. It is interesting to compare the total world 'installed capacity' of human muscle power with that installed in the engines of the U.S.

motor car. If we take the world population of 3.8 billion people and assume that half of these are capable of manual work at a rate of 1/20 h.p., then this works out to a total of one hundred million h.p. The corresponding figure for the U.S. cars is 20,000 million h.p., or two hundred times as much.

Power from the wind and from moving water has been employed since the dawn of civilisation. Coal became increasingly used from the latter half of the eighteenth century and provided the energy needs of the first industrial revolution, initially in Europe and later elsewhere, particularly in the U.S.A. More recently, nuclear power has been developed and is beginning to make a contribution. Of course, the whole energy basis of life depends on solar energy, and the amount of energy falling on the earth's surface is about ten thousand times our current total energy use. Traditionally we have used the solar energy in agriculture for the growth of vegetation, but more recently other uses of solar energy are receiving increasing interest.

In considering the world energy situation we can divide the world into four categories: energy rich, developed (e.g. U.S.A.); energy poor, developed (e.g. Japan); energy rich, undeveloped (e.g. the Middle Eastern oil states); and energy poor, undeveloped (e.g. Sudan). It is with this latter group that we will be particularly concerned in this chapter.

Comparison of energy consumption in the U.S.A. with that in the Sudan

We have pointed out that the energy use in the U.S.A. is as much as 24 times that used per capita in countries such as Sudan. It is instructive to compare the form of energy and the ways in which this energy is used in the two countries (see table 6.1). In the case of the Sudan with its largely rural agricultural economy and low total energy use, the energy demand is dominated by the domestic heat component, mainly used for cooking, a demand which is met by firewood and charcoal. Electricity generation is both a small fraction of total energy demand and a very low value per capita. Comparing these percentages with those for the U.S.A. we see that in the U.S.A. electricity generation accounts for 25 per cent of the total energy input; the industrial requirement is a much greater fraction of the total than for the Sudan. The domestic load is an important fraction of the total and of course in absolute terms is much higher per capita

Table 6.1

Sudan			U.S.A.	
Energy Input				
oil		38%	oil	46%
firewood/charcoal		57%	natural gas	31%
bagasse		2.2%	coal	18%
hydro		1.8%	hydro/nuclear	5%
muscle		0.6%		
Energy Use				
direct heat			commercial and domestic	19%
domestic	61%		industrial	26%
industrial	7%		transport	25%
total		68%	electricity generation	25%
transport		20%	non–energy use★	5%
agricultural tractors/pumping		7%		
electricity generation		5%		

★This heading refers to the petrochemical industry.

than that for the Sudan. The U.S.A. dependence on oil and natural gas is brought out by these figures.

Use of energy in rural areas

The principal uses for energy in the rural areas of developing countries are summarised in table 6.2.

Table 6.2

Agricultural machinery, e.g. the two–wheel tractor
Crop processing, milling, grinding, drying
Pumping water for irrigation and domestic use
Desalination
Powering small industrial equipment, lathes, circular saws, grindstones
Electricity generation for lighting and power, schools, clinics, domestic use
Domestic, cooking, water heating, heating, cooling
Refrigeration
Transport, small vehicles and boats.

There are a number of considerations to be taken into account when selecting an energy source for one of these applications; some of the more important are listed below.

(1) Power level, and whether this is for a continuous or discontinuous supply. For example, if the requirement

is for pumping water, then providing that a storage tank can be incorporated in the system one is only interested in pumping the required quantity of water daily, or even over a longer period, and a wind powered pump may be suitable. On the other hand, if a supply of electricity is required to power an educational t.v. set or the lights in an emergency operating theatre, then continuity of supply is essential.

(2) Cost. There are several cost components which must be considered: initial cost, also its amortisation over the lifetime of the plant; running cost, including fuel cost, if applicable, operator costs, frequency and cost of maintenance.

(3) Complexity of operation. This is of course in relation to the local skills available. A simple hand pump may in some circumstances be preferred to an engine-driven pump simply because of the lack of suitably skilled operators.

(4) Maintenance and availability of spares. What has been said under (3) also applies here. The problem of arranging for adequate maintenance and repair of equipment is perhaps the single most important aspect to be considered in the provision of equipment for rural areas. Development literature contains many references to this problem—for example, the estimate by the *New Internationalist Magazine* (February 1975) quoted with reservations by reference 2. This source estimated that of 150,000 boreholes in India, 90,000 may be out of action at any one time due to pump breakdown.

(5) Lifetime, under field conditions.

(6) Suitability for local manufacture.

Sources of energy

The sources of energy are listed in table 6.3.

Muscle power is given first because of its importance in the rural areas of the developing countries. The fossil fuels currently provide almost all the world energy used. Geothermal and tidal energy have little significance at present, though geothermal energy may prove to have value for some large installations in the future. Solar energy, both indirect and direct, has con-

Table 6.3

Sources of Energy

Muscle power	human and animal
Fossil fuels	coal
	oil
	natural gas
Geothermal	
Tidal	
Solar	direct
	indirect
	wind
	wave
	ocean temperature gradients
	hydropower
	vegetation
	firewood, etc.
	fermentation
	pyrolysis
	animal dung (fermentation)
Nuclear	fission
	fusion
	radioactive decay

---

siderable potential. The disadvantages of low power density*
and variability are often not serious in these situations since
power requirements are low and supply continuity may not be
necessary. Wave energy is not likely to be used generally though
there may be local situations which are suitable. An example of
this is given by Bott in a scheme designed for Mauritius.[3]

Nuclear energy sources are included for completeness but are
unlikely to make any effective contribution in the rural areas.
Having listed the sources of energy we should now consider the
various methods for converting these energy forms to the form
required by the particular application. In practice these forms
will be heat, for cooking, etc., mechanical shaft power for
driving machinery and pumps, and lighting. Conversion to
electricity is convenient for transmission, and for use with
motors or lights. The principal practical problem with electricity
is that of storage; batteries are too expensive for all but small-
scale storage requirements. Most rural applications require only
a few h.p. An exception is the generation of electricity for village
use where up to 100 kW may be employed. Whilst vegetation

* Power density is the term used to describe the amount of energy per second reaching each square
metre.

and other organic material may be used directly, by burning, it can also be converted to gaseous or liquid fuels by micro-biological action (fermentation). In the former case, methane and carbon-dioxide are produced and in the latter case there are a number of compounds of which alcohol is the best known.

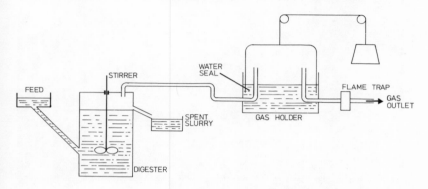

FIG. 6.2   Schematic methane gas generator

The process is simple and figure 6.2 shows a schematic generator. The temperature of the fermenter (or digester) should be maintained at around 35°C, since if it is allowed to drop the microbial action will fall drastically, while higher temperatures will kill off the bacteria. Both the acidity of the contents of the digester and the carbon-to-nitrogen ratio of the contents are important. A properly designed and operated digester will pro-duce a daily output of bio-gas of roughly its own volume. The conversion will be approximately 250 litres of bio-gas* for one kilogram of dry dung, and about 450 litres for one kilogram of dry vegetable matter. Figure 6.3 shows a methane generator.

The apparent simplicity of methane† generation has misled a number of people who have built generators (and worse, pub-lished the details of their designs) without paying proper atten-tion to the required operating conditions, the result being failure. A great deal of publicity has been given to the Indian biogas project which has itself not lived up to the predicted success. Nevertheless, methane generation is an important potential source of fuel for lighting, heating, cooking, and as a fuel for

*1 l of bio-gas has a calorific value of about 24,000 J.
† Methane is an explosive gas and care should be taken in its production and use; flame traps should be fitted. If the outlet to the generator becomes blocked the pressure rise can become sufficient to burst the containing vessel. One of my research students had the misfortune to explode a generator in this way; though fortunately not hurt, he was covered by the contents of the vessel and since the feedstock was animal dung the result was startling and unforgettable.

Fig. 6.3   Methane digester designed by Hutchinson (Thiko, Kenya) *(Photo Peter Fraenkel)*

internal combustion engines; more work on design and implementation is needed. Already successful plants are in operation in various parts of the world.

The fermentation of soluble starches and sugars to ethyl alcohol has been known and practised from the earliest civilisations; there is some interest in the large-scale use of this process for the production of alcohol as a fuel. A report from Brazil suggests that it is the Brazilians' hope to produce as much as 10 per cent of their liquid fuel requirement by the fermentation of cassava by 1980. Of course, the concentration of the alcohol produced by fermentation is low, about 10–15 per cent by volume, so that it is necessary to separate the alcohol from the water by distillation. Fermentation methods can be operated on a small scale and, when properly designed, do not require any high skill in use. They are particularly suited to Appropriate Technology-scale activities.

Destructive distillation of wood (pyrolysis) was an extremely important source of industrial chemicals until the 1920s. In addition to charcoal, a large number of valuable chemicals can be extracted. With the advent of the petrochemical industry, interest in the older method waned and the process is no longer used.

Charcoal production is still practised widely in developing countries; the process used is to reduce access of air by stacking wood
in a kiln or in a pit which has a lid, then burning the wood. The
process is inefficient and can be improved relatively easily.
However, conversion to charcoal loses the alcohols, tars and
other chemicals which can be used for fuels and other purposes;
reference 4 describes methods suitable for small-scale use.

Table 6.4 lists the main types of energy converter which are
classified under end use. We will first look at sources of heat,
which are mainly needed for domestic purposes. The obvious
and widely used source of domestic heat is firewood or charcoal,
but coal, gas and oil can be used, although oil is mainly reserved
for oil lamps. Traditional cooking uses firewood very ineffi-
ciently and improved mud stoves such as that shown in Figure
6.4 could give an immediate and significant fuel saving. Solar
energy is very suited for the production of low grade heat (that is,
low temperature heat). The maximum solar radiation falling on a
flat surface is approximately 1 kW/m²; in a country such as
Jamaica the average solar radiation falling on one square metre
per year is around 2,000 kW hr. Surprisingly, at the equator the
solar energy is somewhat less than in the tropics. Most of the
developing countries lie within the latitudes 40°N to 40°S so that

Fig.6.4   R.I.I.C. improved
stove (Kanye, Botswana)
(Photo Peter Fraenkel)

Table 6.4   **Main classes of energy converter**

Heat	Direct Burning (Wood, Coal, Oil, Gas)	
	Solar Flat Plate	Water Heating
		Air Heating
		Distillation

Muscle Power   Human and animal

Internal Combustion Engines (Oil, Gas)
   Reciprocating
            Petrol–spark ignition
            Diesel–compression ignition
            Humphrey water piston
   Rotary
            Gas turbine

Heat Engines (any source of heat, solar, oil, gas, coal, wood)
      Vapour (Rankine)
            Reciprocating★ Steam Engine
            Rotating Steam Turbine
      Gas (Stirling)
            Reciprocating★
         (Brayton)
            Rotating Gas Turbine
      Electron Gas
            Thermionic
            Thermoelectric

Electromagnetic Radiation (solar energy)
      Photocells

Hydraulic Engines (water power)
      Wheels, Screws, Buckets
      Turbines

Wind Engines (wind power)
      Windmills
            Horizontal axis
            Vertical axis

Electrical/Mechanical
      Dynamo/Alternator (mechanical shaft power)
      Motor (electricity supply)

★Can be constructed using a water piston.

solar energy is generally available. Solar heat is collected by flat–plate collectors (figure 6.5), which are of simple construction and may be used for water heating, distillation and air heating, the latter for crop drying. A flat–plate heater of 1 m² will give 50 to 80 litres per day at 60°C in these areas. Figure 6.6[5] shows a simple water still; this design produces about 1,000 1/yr m². Where higher temperatures are required it is necessary to concen-

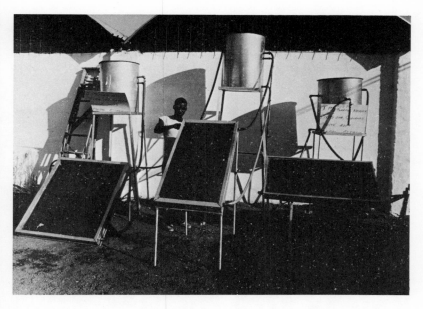

FIG. 6.5   Flat–plate solar water heaters (Gaberone, Botswana) *(Photo Peter Fraenkel)*

FIG. 6.6   R.I.I.C. solar still (Kanye, Botswana) (The brackish water filler is on the left and the distillate emerges from the right.) *(Photo Peter Fraenkel)*

trate the incoming solar energy (every schoolboy knows the burning glass effect); since lenses are expensive, mirrors are chosen. These are made out of thin metal sheet. Figures 6.7 and 6.8 show a design of a solar cooker[6] which employs some concentration.

FIG. 6.7   Solar cooker (Appropriate Technology Development Association, Lucknow, India) *(Photo Madame Marie Farge)*

Muscle power is still the only power source for much of the traditional agriculture, and we have already discussed a number of animal and man–powered pumping systems. One of the most efficient methods of using human muscle power is the bicycle, which provides a great deal of the transport in many developing countries, particularly in the Far East. Wilson[7] states that in normal cycling the cyclist can develop about 0.3 h.p. for a sustained period. The bicycle power train—that is, the seat, pedals and chain drive—can be used for static applications. Wilson has developed this system under the name Dynapod and has adapted it to a number of crop processing applications such as winnowing grain. Figure 6.9 shows a cassava grinder from Nigeria which uses the same principle. Cassava is fed into the hopper and forced against hacksaw blades fixed into a bicycle wheel and moving across the hopper outlet. Figure 6.10 shows a similar arrangement designed by Weir and intended to power a simple wood-working lathe. The flywheel is made from a bicycle wheel which is filled with concrete in order to increase the inertia of the flywheel, that is, to enable it to store more energy.

Internal combustion engines

The spark ignition engine, operating on petrol, gas, kerosene, and more recently bio-gas, and the compression ignition or diesel engine provide a large proportion of the total mechanical power in the developing countries. Diesel-powered electrical generators are also widely used. As pointed out in chapter 5, the principal disadvantages of the commercial internal combustion engines are initial capital cost (often in foreign currency), cost of fuel (again often imported), and particularly the difficulties experienced in arranging maintenance and repair. In spite of these difficulties internal combustion engines will continue to play a major role as energy converters in the foreseeable future. The use of bio-gas as a fuel for internal engines may also be expected to grow. Bio-gas may be used as the sole fuel in a spark ignition engine, and existing engines require only simple modification to change from a liquid fuel such as petrol to bio-gas. In the case of the compression ignition engine, some diesel fuel is also needed (10–20 per cent) in order to obtain ignition.

The Humphrey pump is an internal combustion engine employing a liquid piston and has been described in chapter 5.

The engines so far mentioned are of the reciprocating type, that is, a piston moves up and down in a cylinder, being forced down by the hot, high pressure gases and returned on the exhaust stroke by the energy in the flywheel. The other important geometry is the rotating engine, or turbine; in the turbine the hot, high pressure gases in escaping from the combustion chamber pass a number of blades, causing a shaft to rotate in a similar manner to the action of a windmill. Gas turbines are unlikely to be of use at the relatively low-power levels in which we are interested.

Heat engines

In the internal combustion engine fuel and air are mixed and ignited in the engine.* Another important class of engine is the

---

* Animals including ourselves, are internal combustion engines, employing air to burn the fuel (food), but in this case by a chemical process rather than by explosive ignition.

FIG. 6.8   Improved solar cooker
(Appropriate Technology
Development Association,
Lucknow, India) *(Photo Madame
Marie Farge)*

FIG. 6.9   Pedal-powered cassava grinder *(Photo I.T.D.G.)*

FIG. 6.10 Pedal–powered lathe *(Photo Alex Weir)*

heat engine, which is operated from a furnace or other external sources of heat. The steam engine is an early example of heat engine which, using coal as its fuel, provided the power source for the industrial revolution. In addition to their use in industry steam engines were extensively employed in agriculture, until displaced by the more convenient diesel engine tractor. The old steam engine had much to commend it. It is robust, long–lived and simple to operate and maintain. It is, however, heavy and relatively expensive to build. Steam engines should not be forgotten in the present context and when fired by wood or coal may still have a part to play. Figure 6.11 shows a 4.5 h.p. steam

FIG. 6.11   Ricardo steam engine *(Photo Peter Fraenkel)*

engine designed by Ricardo and constructed from an old internal combustion engine block. More recently, interest has centred on the use of solar energy as the heat source. In order to obtain a sufficiently high temperature to evaporate the water it is usual to concentrate the solar energy by a mirror. One disadvantage of the concentrator over the flat–plate collector is the need to track the sun as it moves across the sky.

Solar energy can be combined with the liquid piston to

provide a very simple pump design. The earliest liquid piston pump, the Savery Pump, was invented in 1698 by Captain Savery to pump water from mines and it employed steam raised by burning coal. This concept has now been revived and combined with solar heating, by Rao,[8] and a number of proposals for solar-powered heat engines using liquid pistons have appeared in the literature. It is not necessary to use water as the working fluid and there may be some advantage in using a fluid with a lower boiling point in the solar heating applications. French workers have developed two such designs, one employing a reciprocating piston and the other a small turbine.[9] Of course, in these designs the vapour must be conserved by condensing as it leaves the expander and then must be returned to the boiler.

So far we have discussed engines which use the evaporation of a liquid to produce a high pressure vapour which in turn drives the piston or turbine. We could instead use a gas — for example, air. In the Stirling engine, a gas is heated and allowed to expand; in so doing it pushes a piston out, doing work. The hot gas is then transferred to a cold cylinder where it is compressed; the work to compress the cold gas is less than the work obtained by expanding the hot gas, and there is thus a net work output. The cold compressed gas is returned to the hot cylinder and the process repeated. Stirling engines have the efficiency and other characteristics of a diesel engine but with the added advantages of accepting any fuel, and, since they are sealed units, they will operate for long periods without maintenance. The major objection to the Stirling engine for our purpose is the cost, probably 20 per cent higher than the equivalent diesel, and the complexity of manufacture making it difficult or impossible to manufacture locally. An interesting development is the Fluidyne[10] which employs a Stirling cycle in association with a liquid piston for pumping purposes. The Fluidyne will be of low cost and suitable for local manufacture.

We should mention another form of heat engine, the thermocouple, which consists of two different metal wires whose two junctions are maintained at different temperatures. It is found that under these circumstances an electric current will flow round the circuit. Thermocouples are useful for generating small amounts of electric power (a few watts) but are unlikely to be of general use in developing countries for reasons of high cost and low efficiency. The thermionic generator requires very high temperatures (greater than $1,200°C$) so is even less likely to be of any use for these applications.

Electromagnetic radiation — the photocell

The photocell is familiar as the electric power generator which operates the photographic exposure meter. Photo–electric generators based on silicon have been developed with an efficiency of around 14 per cent and are extensively used in the space programme. However, their cost is far too high for terrestrial applications. More recently developments have been reported which claim that the costs can be appreciably reduced; if this proves to be possible then the photoconverter will need to be reassessed for such applications as electric lighting, and even pumping. Wijewardene has applied photocells to crop spraying.

Hydraulic engines

Water wheels have been used since the time of the Romans and possibly earlier. They may be used in conjunction with dynamos for the generation of electricity or for milling and grinding and other uses. There are a number of designs of both wheels and simple turbines available.

A simple design of water wheel/electric generator based on the bicycle wheel has been developed in Colombia. This has exposed a problem associated with the use of bicycle parts. In Appropriate Technology, one tries to use readily available materials and of course bicycle parts come immediately to mind. Unfortunately, whilst it is easy to convert bicycle parts, for example into small water turbines, it is equally easy to convert them back again, and a small local industry for this purpose has grown up using stolen generators. It is unlikely that the local entrepreneur will be able to reconvert the concrete-filled wheels of figure 6.9. Figure 6.12 shows a water turbine producing 1 kW. One feature of this design is the simple power conditioning gear, an electronic box which maintains a constant voltage output in with variable load and input power.

Wind power converters

We can divide windmills into two types, the horizontal axis machines in which the vanes rotate about a horizontal axis, and the vertical axis machines. Examples of both type are shown in

FIG. 6.12   1 kW water
turbine (propeller is
shown in front
of housing)
(Photo Peter Fraenkel)

figure 6.13. A disadvantage of the conventional (horizontal axis)
windmills is that they must be positioned always to face the
wind. Windmills intended for pumping applications are required
to rotate relatively slowly and normally have a large number of
vanes, whereas windmills intended for electricity generation
rotate rapidly and have a small number of vanes (two or three).
One problem in electricity generation is the need for storage.
This is usually supplied by batteries (accumulators) and it limits
the output of the mills to a few hundred watts. Much larger
electricity generating mills have been built and are currently
being intensively studied; these are usually incorporated in the
electricity grid.

Windmills for water pumping have been used extensively in

FIG. 6.13   Vertical and horizontal axis windmills at Reading University *(Photo Peter Fraenkel)*

FIG. 6.14   Wrecked windmill (Sudan) *(Photo Stuart Wilson)*

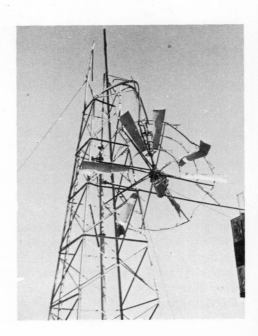

FIG. 6.15 I.T.D.G. windmill, Reading University *(Photo Peter Fraenkel)*

the U.S.A., Australia and elsewhere for many years. Over the past few decades they have to some extent been replaced by internal combustion or electric motor-driven pumps. More recently, with the increase in fuel costs, interest in wind powered pumps has revived.

Windmills have not been used on the same scale in the developing countries, for several reasons:

(1)   The relatively high capital cost of imported mills when compared to the cost of the equivalent diesel engine.

(2)   The lack of dealer and servicing facilities, resulting in poor maintenance and early failure. Parkes[11] quotes for Tanzania only 30 mills of which as few as 10 are operational. This compares with 185,000 operating in South Africa. Similar high failure rates are reported for other developing countries. Figure 6.14 (from the Sudan) shows what happens when the gear box is not regularly refilled with oil.

(3)   Existing commercial designs, though well proven, are old designs and are both material intensive and expensive to build.

The I.T.D.G. windmill was designed specifically for local manufacture and maintenance. Figure 6.15 shows the first prototype built in the Engineering Department at Reading University. At present a programme is in hand to provide this design for local manufacture in a number of developing countries.

7

Services — Medicine, Building, Transport

Medical care and equipment

Figures were given in chapter 2 for the average number of people per doctor in the developing countries; it was seen that the position is very much worse than in the developed countries. The actual situation is in fact worse than these figures would suggest since most doctors tend to live in the cities and towns, and very large numbers of people living in rural areas do not have access to any trained medical assistance at all. It is said that every half minute 100 children will be born in the developing world* and of these 20 will die in the first year; of the remaining 80 some 60 will not receive any medical attention during childhood. Some disturbing figures are quoted by Harrison[1] in a recent article dealing with medical care in Colombia and Peru. He states that in the Puno area of Peru [an area of 72,382 sq km with a population of 780,000 (1972)] 33 of the 42 doctors, all of the 8 dentists, and 43 of the 44 nurses live in the three small towns. In Colombia 70 per cent of the doctors and 86 per cent of the nurses live in department capitals, and 32 per cent of the doctors live in Bogota itself. Children are particularly vulnerable and the infant mortality rate for Colombia is between 60 and 70 per 1,000 live births, compared with a figure of 18 for the U.K. In the Puno region of Peru it is estimated as between 200 and 500 per 1,000.

One method of providing at least some medical care is by training medical auxiliaries or paramedics. The best known of these schemes is the Chinese 'barefoot doctor' development.[2,3] Field workers are selected and given a three months' course of basic training, which may be supplemented by further short

* If we assume the population of the developing world to be 2,800 million and a birth rate of 4 per cent, this gives 112 million births per year. There are 525.6 thousand minutes in a year, which works out at about 213 births a minute, so the figure is roughly correct.

133

courses in later years and by in-service training by visiting medical teams. The barefoot doctor provides simple diagnosis and prescribes treatment in both traditional and modern medicine. Examples of the former include herbal treatments and acupuncture. His main function is in prevention, by inoculation, health education, and hygiene, and in pest eradication. The barefoot doctors form the lowest layer of a medical pyramid, and are closely integrated with the higher levels of medical care through clinics, local hospitals and specialist area hospitals. A notable feature of the barefoot doctor scheme is the large numbers involved; a unit of a commune of perhaps 50 families (200 people) could have two or even three barefoot doctors attached to them. The medical role forms only part of the barefoot doctor's duties and for the remainder of his time he continues to work in the fields in the usual way. They are not, of course, barefooted when practising medicine—the term derives from their other work which may involve wading in the paddy fields.

Similar approaches to the provision of intermediate medical care have been started in other countries, notably Tanzania, Peru and Colombia. I.T.D.G. has its own medical panel who have been investigating proposals for the training of such medical auxiliaries.[4]

The high birth rate in the developing countries has been referred to elsewhere, and figure 2.11 shows the effect of this on the population structure for India. There seems to be evidence that the development of a rural medical service is accompanied by some limitation of families. This is partly due to the dissemination of information on birth control techniques, but is also connected with the reduction in the normally high rate of child mortality. In countries where there are few welfare services, children represent security for old age, and a large family is considered essential to ensure that several children survive. The limitation of the world population is the most urgent development problem. This goal is unlikely to be achieved by exhortation, nor simply by the provision of birth control techniques; it will only result when the bulk of the world population feel sufficiently secure for themselves to wish to limit their families.

A very interesting and potentially important development for the prevention of diseases such as cholera, dysentery and typhoid from food infected by handling is reported from the Sudan.[5] The device is called a *tabag* and consists of a small straw or palm leaf tray, fitted with a transparent conical plastic lid which is arranged to fit closely around the rim of the tray, and placed in the sun.

Food such as bread, pies or fruit is placed in the *tabag* for approximately one hour. It is found that the inside temperature rises to 55°C within about 25 minutes, and most intestinal bacteria die if exposed to temperatures of around 54° to 56°C. Tests showed that the count was considerably reduced after 45 minutes' exposure in the *tabag*; there was complete inactivation after one hour, and after 1¼ hours no organisms could be recovered at all. This device is cheap and simple and could be used to reduce the chance of infection at religious and holiday gatherings and to protect travellers such as pilgrims who buy food from wayside stalls. It is particularly useful for treating foods such as bread which cannot easily be washed.

Figure 7.1 shows some examples of simple medical equipment made from plastic drain pipes. This work was carried out in Nigeria and required very few tools and used local unskilled labour.[6]

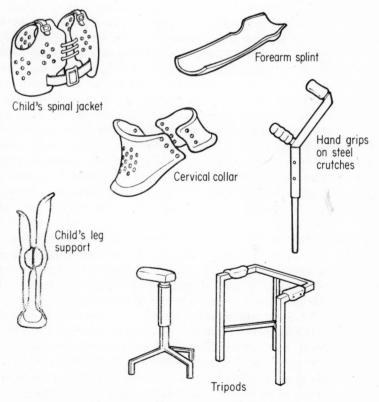

Child's spinal jacket

Forearm splint

Cervical collar

Hand grips on steel crutches

Child's leg support

Tripods

Fig. 7.1   Medical aids made from plastic drain pipe *(Reference 6)*

In Chapter 1 an example of hospital equipment was described. It was pointed out that since this equipment was designed and built locally it took account of local conditions and also could be readily maintained and repaired. Further designs are given in reference 7. Some examples of surgical instruments which are made by small industry in India are described in reference 8.

Building

A building must satisfy a number of requirements; it should provide shelter from the sun and rain, and protection from the wind; it should reduce the effects of the sun in the day time and the chill in the night. In addition to these requirements, which are essentially concerned with the provision of an acceptable environment for the occupants, the building should also provide protection from insects, animals, ground water, and give privacy and security for possessions, and in the case of town dwellers for the person also. Parry suggests[9] that these requirements are best met by a dry cave; he points out that caves maintain moderate ambient temperatures, have low maintenance costs, can be maintained in a hygienic state, and are easily secured against thieves and animals. However, there is a limited supply of suitable sites and we cannot all live in caves. The Chinese tried a somewhat similar approach by constructing a small town in which all the houses are below ground; in this way they take advantage of the virtually constant temperature which exists below the surface. We will first discuss some traditional building solutions and then go on to describe a number of Alternative Technology initiatives in the building field.

Climatic and other conditions vary so widely that there can be no single building solution. Figure 7.2 shows a number of different traditional housing types from various parts of the world. It contrasts a typical traditional house in the humid tropics, such as Malaysia, with a house from a hot arid country as in the Middle East. In the first example the structure is of light weight and is arranged so that the air can move freely through it. It is built on stilts to lift the floor well above ground water level. In contrast the second house has massive walls and roof which protect the occupants from the sun in the day, and at the same time store heat to maintain a reasonable temperature during the cold night. The 'modern' solution, a box with walls constructed

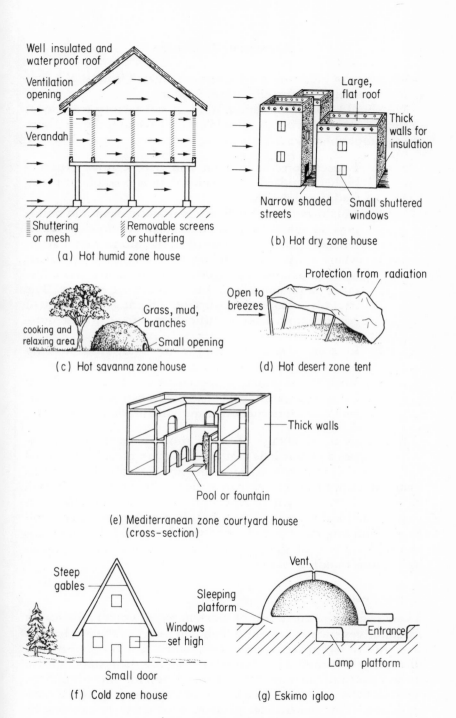

Well insulated and waterproof roof

Ventilation opening

Verandah

Shuttering or mesh

Removable screens or shuttering

(a) Hot humid zone house

Large, flat roof

Thick walls for insulation

Narrow shaded streets

Small shuttered windows

(b) Hot dry zone house

cooking and relaxing area

Grass, mud, branches

Small opening

(c) Hot savanna zone house

Protection from radiation

Open to breezes

(d) Hot desert zone tent

Thick walls

Pool or fountain

(e) Mediterranean zone courtyard house (cross-section)

Steep gables

Windows set high

Small door

(f) Cold zone house

Vent

Sleeping platform

Entrance

Lamp platform

(g) Eskimo igloo

Fig. 7.2 Traditional housing types *(Reference 10)*

from concrete blocks and a roof of corrugated iron, is a design quite unsuited for many climatic conditions.

We should ask the question Why, if satisfactory traditional designs have been evolved, should we try to change them? There are a number of reasons for wanting to do this.

(1)   Expectations have risen and people now expect a higher standard of amenity than is possible in many traditional designs.

(2)   Much traditional material is impermanent and buildings in which it is used require continuous maintenance. (This is not true of all traditional materials.)

(3)   Many traditional building materials, such as leaves, bark, unpainted wood, etc., are subject to insect infestation and hence can be a source of disease. The irregular nature of the surfaces and the general form of construction make it difficult to screen against mosquitoes.

(4)   In some areas the supply of traditional building materials, particularly wood, is becoming scarce.

(5)   With the increase in specialisation, labour costs are rising and traditional construction can be more expensive than construction employing modern materials.

A serious disadvantage of using modern materials such as cement, glass and plastic is that most of them must be imported and hence will require scarce foreign currency; also because this material is imported, rather than made locally, potential jobs will have been lost. One aim of Appropriate Technology building is to provide local sources of new building materials, which will enable building designs to be evolved that avoid the disadvantages of traditional designs, at the same time providing an indigenous building industry.

Building materials [11]

There has been a great deal of work in recent years in the search for additives such as cement which will improve the mechanical strength of compacted soil. This work has been primarily directed at improving road construction, but the results can also be applied to the making of bricks from soil for house building. Traditionally sun–dried bricks are widely used in house construction. The mechanical properties of these bricks can be considerably improved by adding a small proportion of cement (around 5–8 per cent); this amount of cement is very much less than would be required for concrete blocks which are the alternative and expensive solution. Soil stabilised bricks have been developed in a number of countries. Figure 7.3 shows the Cinva ram developed by the Village Technology Demonstration Centre at Karen, near Nairobi, Kenya; the ram is used in the manufacture of soil stabilised bricks.

FIG. 7.3   Cinva ram (Village Technology Centre, Kenya) *(Photo UNICEF)*

Fired–clay bricks have very much improved mechanical properties over the sun–dried bricks. Where a suitable source of clay exists together with a supply of fuel these bricks can be produced on a small scale. Parry[9] describes such a low–cost brick-making factory which he set up in Ghana. The first factory at Asoka produces about 10,000 bricks a week and employs 26 men from the village. The significant fact about this factory is that it cost only $20,000 to set up, or about $800 per work place. These figures should be compared with a cost of $40,000 per work place for a conventional brick-making factory. Parry states that, in cost and quality, the bricks are comparable to those from a conventional plant.

Traditionally grass, leaves and reeds have been popular choices for roof materials. Fired–clay tiles provide an alternative which can be made on a small scale but have the disadvantage of requiring a relatively massive timber support structure. The need for the latter can be avoided by a dome or vault–type construction which is common in the Middle East and elsewhere. Sheeting made from corrugated iron or asbestos cement is now widely used. It is easy to fix but has poor thermal insulation properties. Waste paper, grass, straw, wood wool and coir fibre can also be used to make corrugated roofing sheets.

Windows and doors with wooden frames and installed in wooden surrounds are expensive. Baker,[12] in Kerala, India, has revived the use of open brickwork, known locally as *jali*, which provides both light and air and eliminates the need for conventional windows (figure 7.4). Baker has also developed windows and doors which have simple dowel hinges and which require only lintels for their support (figure 7.5).

FIG. 7.4　*Jali* open brickwork (Kerala, India) *(Photo by Jack Skeel. Courtesy I.T.D.G. Reference 12)*

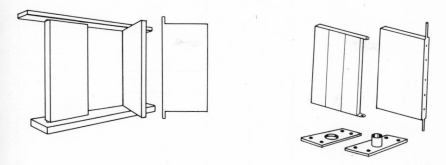

FIG. 7.5   Dowel hinges *(Reference 12)*

Small-scale lime kilns have a long history, and simple designs suitable for local use are available. Until the invention of Portland cement in the nineteenth century, most European masonry buildings used mortar made from slaked lime, sand and fibre such as hair to act as a binder. Portland cement requires very high temperatures in its manufacture (1,450°C) and is usually made in large industrial plants, but it is possible to produce this material on a small scale and work on such a plant, capable of an output of 2–30 tonnes a day, has been developed by C.S.I.R., Assam, India.[13] The capital cost ranges from U.S. $18,000 for 2 tonnes per day to U.S. $180,000 for 30 tonnes per day. The product cost is U.S. $22 per tonne which is not significantly higher than the production cost from large-scale plant. The small plants can be built locally and thus eliminate the costs of transport which can be high. Pozzolana cement has a long historical tradition and could provide an alternative to Portland cement in domestic and other small building construction. This cement is made by grinding fired tiles or bricks to a fine powder, mixing with slaked lime and adding water. Pozzolana cement was extensively used by the Romans and is still used in some countries, for example, India, where it is known by the local name of *surkhi*.

Heating and cooling

Careful design of a building can contribute a great deal to the maintenance of a comfortable environment; this method is known to architects as 'passive system design' to distinguish it from mechanical systems of heating or cooling. There is much to learn from traditional practices. Figure 7.6 shows a traditional

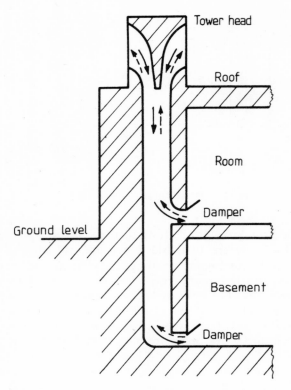

FIG. 7.6   Operation of an air tower in summer *(Reference 13)*

form of building construction in central Iran, a design known as
an air tower.[14] This area of Iran has a desert climate with clear
night skies and a high exposure to the sun all the year round. The
air tower provides summer cooling and almost all the necessary
winter heating. The local buildings have flat roofs, thick walls
and ceilings and hence store a large quantity of heat (that is, they
have high thermal capacity). Cooling during the hot, dry,
summer months is achieved as follows. At night the windows
and dampers are opened and the towers operate as chimneys
drawing in the cold night air to cool the structure. Air passing
through the basement is further cooled by picking up moisture
which has diffused in through the ground. In this way the tower
walls are cooled. During the day, hot dry air is drawn down the
tower where it is cooled before entering the rooms. Air towers
are built with openings on several sides to enable the afternoon
wind to be used to provide additional circulation.

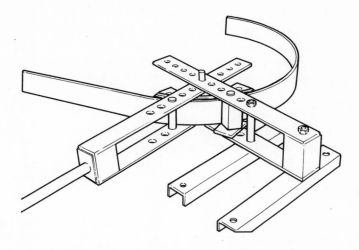

FIG. 7.7 I.T.D.G. metal bending machine *(Reference 14)*

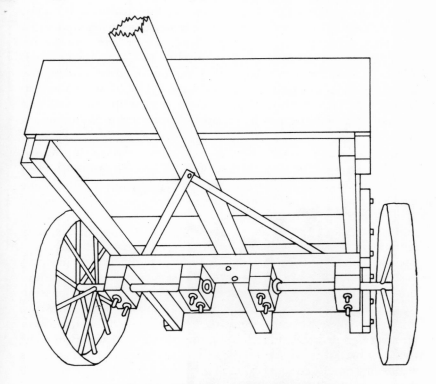

FIG. 7.8 'Wanochi' ox cart *(Reference 15)*

Other services

Sanitary fittings can easily be made on a small scale. The provision of a water seal is a great improvement, and examples of such fittings have been manufactured to designs produced by the Planning Research and Action Centre at Lucknow, India.

The Tema house

An interesting example of A.T. housing is given by the Tema development described in chapter 1.

Transport

Transport is needed for both goods and people. Domestic and farm work can often be greatly assisted by the provision of simple, two-wheeled hand carts for carrying water vessels or farm produce. Wood may be used to replace most of the metal parts used in the equivalent commercially produced vehicles. Two-wheel animal-drawn carts can easily be constructed and a number of designs are available.[15] Scrap car or lorry back-axles and wheels are obviously very suitable and are widely used. Where wheels are not available they can be constructed using the metal bending machine developed by I.T.D.G.[16] and shown in figure 7.7. This method requires only steel strips and rods and the use of a welding torch. Figure 7.8 shows the 'Wanochi' ox

FIG. 7.9   Construction of concrete barge (Sudan) *(Photo I.T.D.G.)*

cart developed by the Reverend Swenson in Tanzania. This cart makes use of oil-soaked wooden bearings.

Boats are a common means of transport in many developing countries. Reinforced concrete or resin-bonded fibre-glass offer alternatives to traditional boatbuilding methods; they require the same skills and are labour-intensive and do not require special equipment. Dickinson[17] reports on the development of reinforced concrete boats in China, using chicken wire to shape the hull and concrete as the filler. Such boats can readily be repaired should they suffer damage. Figure 7.9 shows the construction of a concrete barge in the Sudan (a project being carried out by I.T.D.G.). Figure 7.10 shows one of the completed barges. Fibre-glass reinforcement with resin-bonding and reinforced concrete are suitable for other village applications, such as the construction of rat-proof storage bins and water storage tanks.

FIG. 7.10   The completed barge
(*Photo I.T.D.G.*)

8

Small Industries in Rural Areas

There are few ways in which a man can be more innocently employed than in getting money.

Samuel Johnson, 1775

There are several strong reasons for encouraging the development of a rural industry which manufactures equipment specifically designed to satisfy local requirements, and which makes use of local skills and materials. These include the removal of the drain on foreign currency resulting from the import of goods from outside the country, and the problems arising in the servicing, maintenance and repair of such equipment. Spares are often difficult to obtain and maintenance skills are lacking. Examples of the latter have been given in chapter 6, in the case of windmills in Tanzania and the Sudan. Also, though such equipment will have been well designed for its original purpose, it does not necessarily satisfy all local needs and practices. Technologies do not readily transplant. The creation of work places is of immediate value in opposing the drift from the rural areas by providing new job opportunities. Perhaps the greatest value of the growth of such industries is in the resulting education and training of the rural people, thus providing a means for developing their skills and experience on a continuing basis.

Particular problems of small industry and the small entrepreneur

At the outset it should be said that entrepreneurs are born and not made. What is required is an environment which provides them with encouragement and support in order to enable their activities to prosper. Their motivation is of course profit, and Dr Johnson's remark quoted at the beginning of the chapter is particularly applicable if we replace 'innocently by 'single-mindedly'.★ It is found that the reasons why entrepreneurs encounter difficulties in building up their businesses fall under three headings:

(1) Lack of management skills, including organisation, financial control, planning and marketing.
(2) Shortage of capital.
(3) Technical problems and lack of expertise.

We will treat these three items in turn. Firstly there are the problems of management. A typical case might be that of a small one- or two-man traditional carpentry business. The craftsmen would normally serve their immediate local area and often the work will be commissioned on a one–off basis. The owner will probably have little appreciation of estimating, pricing, work planning, ordering of materials, etc. He may not keep even the simplest of accounts and have little idea of whether he is making a profit and certainly not know which lines are most profitable. These small businesses may well have craftsmen of high skill or skill potential; they will, however, be restricted to traditional practices, tools and locally available traditional materials. In order to expand such a business it is necessary to seek a wider market. This will have a number of consequences; for example, new lines will have to be developed and it will be necessary to introduce simple quality control of the products. Such considerations and changes will require help and guidance from outside the firm.

Secondly there is the problem of the availability of capital for expansion of the business, new building, new plant or just working capital. Most developing countries have finance and banking arrangements set up for the development of small industry. In practice, however, the situation is not so easy. Before lending money a bank must assure itself that the recipient has some financial credibility and it is often difficult for a bank clerk, who does not generally have suitable industrial experience, to assess the potential of the client and the technical and commercial viability of a new project. The result is that frequently the request is turned down. What is required is for the case to be properly assembled and presented to the bank in terms it will understand.

Thirdly there is the question of technical skills; the local craftsman may be aware of a method or a piece of equipment but

★To those people who find the profit motive distasteful, it must be said that it is difficult in this situation to find any alternative motivation as powerful and effective. The spirit of community service which apparently exists in China is the most notable exception to this statement.

not of its suitability for his particular job. He needs answers to such questions as how much better will a new technique be than the traditional method? Will the use of the new technique justify the cost of the equipment? How will he obtain the training to use the new equipment? To take an example, let us suppose that a traditional village blacksmith is considering the purchase of a gas welding set. All three types of difficulty arise. How will he obtain the money? How will the loan be repaid? Where will he buy the equipment, and to what specification? Finally, who will show him how to use it in his own business? Sometimes the technical level of the problem can be surprisingly simple; for example, in a small foundry in South Vietnam we found a high wastage rate which was readily improved by a simple modification to the mould. In the same area we visited a one-man firm engaged in the production of pump pistons and impellers from light alloy using very simple home made equipment. The feed material was war surplus made to a high military standard and available in quantity. The quality of the product was quite acceptable but the owner wished to increase his output and standard by installing a commercial die-casting plant. He was fully aware of the commercial range of equipment available and had carefully considered his needs before making a choice. His only problem turned out to be the one mentioned previously, that of convincing the bank, several hundred miles away in the capital city, of his financial viability and the security of their investment.

But not all entrepreneurs need help

One outstanding example of the successful entrepreneur is found in the case of the Winner Engineer Company in Bangkok, in which they modified a technology to suit local conditions and has built up a thriving market. The company developed and now manufactures a small two-wheeled tractor. This tractor, which is similar to the imported Honda version, sells for only half the price. These machines are powered by engines which, with the exception of the German imported carburettors, are entirely locally made. The engines themselves are sold in six different models, ranging from 3 h.p. to 30 h.p., and sell at around 15–30 per cent less than the imported Japanese equivalent. The engines can be used to power boats and pumps in addition to tractors. Apart from ironfounding, the manufacture and assembly is carried out almost entirely by female labour. The tractors are robust and durable, and spare parts and servicing are readily available; above all they are cheap and because of this they have severely reduced the sales of their Japanese competitor; they are also exported to Malaysia and the Philippines. It is salutary for the experts like myself to note that this manufacturer does not employ graduates, has not received technical or other advice and encouragement from experts, nor has he received financial assistance. It is estimated that there are around one hundred small firms of this type in the Bangkok area engaged in the production of farm machinery and implements.

In general, however, the entrepreneur requires help in one or more of the three areas — capital, management and technical skills — and a number of organisational solutions to his problems are available.

National and Regional organisations

CoSIRA, U.K.[1]

In the U.K. a government-supported organisation, CoSIRA, has been in existence for this purpose for the past forty years. CoSIRA, or the Council for Small Industries in Rural Areas, was set up to assist small firms in the rural areas of England. It is restricted to businesses employing not more than 20 skilled persons and situated in country areas or towns of a population less than 10,000. Agriculture, horticulture, and the retail trades are excluded. The council offers advisory services, including management advice, technical advice, experimental facilities, and specialist training courses. It also publishes a register of specialist engineering firms which is supplied to larger industrial organisations to make them aware of the availability of these skills in the rural areas. The Council operates an Industrial Loan Fund for the purchase of buildings or plant and for working capital. A network of full-time local advisers is maintained, and voluntary local committees consisting of local business and professional men provide advice based on experience and local knowledge. The variety of activities helped by CoSIRA ranges from traditional crafts such as thatching, blacksmithing and pottery to the most modern high technology including the manufacture of high vacuum equipment and computer core stores.

National Sharecroppers Fund, U.S.A.[2]

The National Sharecroppers Fund was incorporated in 1943

primarily to assist small farmers and sharecroppers in the southern American states. Its objectives seem to be somewhat similar to those of CoSIRA, but unlike CoSIRA its orientation is primarily agricultural. Through the Rural Advancement Fund it provides financial, technical and organisational assistance to a variety of projects including agricultural, seafood, and craft cooperatives, child development centres, health programmes, and rural housing projects. N.S.F. has also set up an experimental farm and training centre. It is hoped that the centre will become a national and, in time, an international source of knowledge and technical assistance for people in rural areas.

Regional organisations in developing countries

In many developing countries the local university or a local research institute may represent the only source of physical science or biological science expertise, or engineering know-how. The sociology and economics departments of universities can also provide valuable assistance in local development programmes. Other local resources include banking and the local branch of the ministry concerned with small industry development. There is a need for a centre to coordinate these resources and to act as a communication and interpretative link between them and the small entrepreneur. A centre should also provide some specialist knowledge and workshop facilities. The T.C.C. at Kumasi, Ghana, is a good example of such a centre (see below).

Figure 8.1 illustrates the interactions of a unit such as this with other organisations. As far as the university is concerned, this is not a one-way traffic, since projects arise which are useful in providing projects for undergraduates and M.Sc. and Ph.D. students. This sort of involvement with industry by a university also goes some way towards a rebuttal of the 'ivory-tower' charge which is often made.

The notes below are based on a proposal to set up such a unit on a university campus in West Africa, and outline the activities and terms of reference of the unit.

(1) To provide an advisory/consultancy service for industry and government bodies, acting as a communication link between industry and university staff.

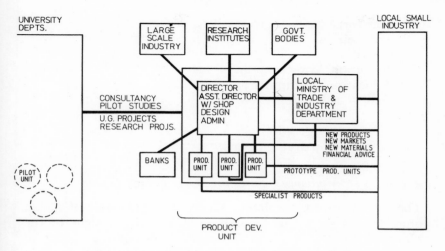

FIG. 8.1   Typical interactions of a product development unit

(2)   To set up small production units to be operated as commercial enterprises and ultimately handed over to private industry.

(3)   To develop new products and introduce them to local industry.

(4)   To provide an advisory service on the purchase of equipment, and help with servicing the equipment (including the provision of special replacement parts) if it is of a complex nature.

(5)   To act as a supplier of specialised components if there is local difficulty in obtaining such components.

The unit would receive income from both sales/services and products and the aim should be to become broadly self-supporting. It is in fact essential to provide services on a commercial basis in order to maintain commercial realities on the part of the recipients, otherwise they become involved in an artificial situation and a continuing dependence.

Possible areas of activity would include

   Ceramics
   Stabilised soil bricks
   Low-cost buildings
   Soap manufacture
   Essential oils
   Rubber

Timber: extraction, treatment and use
Palm oil
Textiles
Irrigation, low-cost pumps
Educational equipment
Food—fish farming
          canning
          refrigeration
          crop drying (solar).

Technology Consultancy Centre (T.C.C.), Kumasi, Ghana[3]

The Technology Consultancy Centre is an example of such a unit. It was set up in January 1972 as a department of the University of Science and Technology, Kumasi, in order to strengthen links between the University and industry and to help promote industrial development. The Centre was intended to acts as a link between professionally qualified personnel in the University and particularly the local entrepreneurs. The permanent staff is small in number, initially consisting only of the Director, Assistant Director, one or two research assistants and clerical support. More recently their numbers have been increased by the addition of more research assistants who act as trainee project managers on specific projects. An early success of the Centre was in the development of spider glue.

Mr S. K. Baffoe was a small manufacturer of cassava starch which he sold to local laundries. In March 1972 he approached the Centre with a request for assistance in the making of a good quality paper glue. Paper glue can readily be made from cassava starch and this is done in many local schools for their own use, but this glue deteriorates rapidly and requires the addition of a non-toxic fungicide if it is to be kept for more than a few days, and a further disadvantage of the simple glue is that it cannot be rewetted. These problems were solved by a member of the chemistry department of the University, Dr D. N. S. Rao, and Mr Baffoe carried out preliminary trials successfully. The University assisted him in his negotiations with the bank to obtain the necessary loan (about £1,000) to enable him to proceed. Progress subsequently has been remarkable; Mr Baffoe now has seven employees and effectively supplies the Ghanaian market, and is also looking into export possibilities (figures 8.2 and 8.3). Of course not all applicants for help from the Centre have the energy, enthusiasm and initiative of Mr Baffoe, but this example

FIG. 8.2   Spider glue factory (Kumasi, Ghana)

FIG. 8.3   Cassava drying for spider glue

shows what can be achieved with the right man and the right support.

Another initiative shown by the Centre has been to secure a contract from a manufacturer of hand tools in Europe for the supply of a large quantity of wooden handles. There is a wood-working tradition in the area around Kumasi but the one/two-man type businesses would neither be able to supply the quality nor, as an individual business, the numbers required. By break-ing the order down into smaller subcontracts and by providing help on simple jigs it was possible to meet both the quantity and quality requirements.

In addition to its work with the local entrepreneurs the Centre has set up a number of production units. The subjects include soap-making (figure 8.4), bolt-making, broadloom-weaving and others. In the case of soap-making a small factory has been set up some miles from the University campus. With this exception the production units have not resulted in the work being taken up by local business, probably due to the higher capital cost of the work.

A further University activity has been the provision of advice to government departments and larger industry.

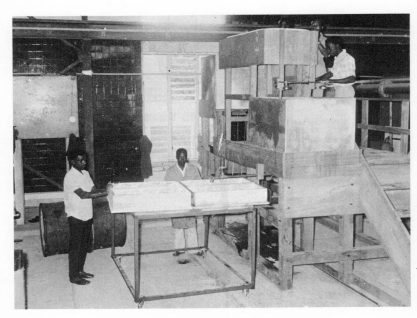

FIG. 8.4   Soap making plant *(Photo I.T.D.G.)*

Whilst it is too early to draw conclusions from the results of the Centre, the preliminary results give grounds for encouragement.

Appropriate Technology Development Association (A. T.D.A.), Lucknow, India[4]

The Association was set up in 1976, its main objectives including the development of small-scale production technologies, pilot plant, carrying out surveys to identify needs, and the dissemination of information by lectures, seminars and publications. One example of the latter, the *A. T.D.A. Handbook*, has already been referred to in previous chapters.

A.T.D.A. concentrates on the development of small industries which should be competitive with large-scale technology and not protected by subsidy or other means. One example of the work of the association is the development of a Portland cement unit which has an output of 35 to 50 tons per day. Large-scale production plant employs a horizontal kiln, with an output of typically 600 to 800 tons per day; this design does not scale well to low output levels and a vertical kiln has been chosen. These small plants can be widely dispersed to be near the consumer markets, avoiding the need for expensive transport.

Other centres of a similar nature to the above have been set up, for example the Appropriate Technology Group in Sri Lanka (A.T.G.S.L.).

Individual projects

Technoserve Inc., U.S.A.[2]

Individual projects also have a role and one is reported by Technoserve Inc., a non-profit corporation, set up in New York State in 1969. The objectives, approach and findings are very similar to those of I.T.D.G. They report[4] on a project in Ghana in 1974. The job was to rationalise the practice of an existing small sugar processing plant consisting of crushing cane and marketing juice, and then to evaluate the viability of a plant for the production of raw sugar and its by-products. The new feature was the relatively small plant size required, 20 to 40 tons per day of sugar cane. Commercial practice elsewhere was concerned with very much larger throughput than this and even the Indian development of the small-scale Khandsari plant was designed for

60 tons per day. Literature on the technology of extraction and plant available was studied and a solution found and successfully implemented.

It is instructive to note the preliminary questions and discussions leading up to the plant specification.

(1) What are the ultimate goals and potential of the project in human and economic terms?

(2) What is the appropriate/possible product-mix in financial/economic terms?

(3) What are the appropriate technologies? How much sugar cane/local labour and skills/local capital, is available? What foreign exchange will be required?

(4) What technologies are available and from where?

(5) What is the state of the domestic market?

(6) What is the attitude of the local communities/ government?

(7) What are the financial returns and degree of risk?

Partnership for Productivity (P.f.P.), U.S.A.

This foundation was conceived in the U.S.A. in 1966 and commenced its first field programme in Kenya, in 1970. It has identified management problems as the principal difficulty in the establishment and encouragement of small industry, and in the promotion of trading activities. The workers have identified as weaknesses the lack of skills in management, planning, marketing and organisation. Other aspects mentioned are the requirements for supporting services and education and training. The approach of P.f.P. is described in a book published by I.T.D.G.[5]

I.T.D.G.

The I.T.D.G. windmill described in chapter 6 was designed for adaptation to different local requirements and for suitability for local manufacture. The method adopted is to send out a project officer to the country for a short period. He takes with him plans of the mill and, in association with a local manufacturer, commissions the construction of a prototype machine. This machine is then erected on a test site where there is skilled staff available for obtaining performance data—for example, on a university campus or at a research institute. Based on this experience any necessary design changes can be made and passed on to the original manufacturer who will be licensed for the production of further mills. A similar project is in hand with the Humphrey pump and also with other equipment.

Low cost automation

Automation is usually associated with large-scale, capital intensive, low-labour situations—for example, car manufacture —and at first sight would seem to have no place in the small-scale, low-capital intensive, labour-intensive requirements of rural industry in the developing world. This turns out not always to be the case and there are occasions where some measure of automation is the most appropriate solution. Automation is often regarded as synonomous with feedback control in which an operation can be carried out completely automatically. At the level with which we are concerned here, automation is more or less the same as mechanisation. In a developing country, although there is an abundance of unskilled labour, this is not so for skilled labour, whose shortage is often a serious industrial bottleneck. By simple automation, as in the use of capstan lathes, the skilled personnel can be employed on such operations as tool-setting, and semi-skilled labour used in the actual production; and secondly, the use of simple jigs and fixtures enables standards to be set so that quality control becomes much easier. The conversion to low-cost automation will usually enable existing machinery to be employed and so does not involve costly capital expenditure. In this way the use of low-cost automation will increase job places, provide a better product, and give improved utilisation of existing plant. Low-cost automation is of course not a universal panacea; as Rijn de Groot, a small industries expert, once said to me, 'If the system is a mess and you automate it, you have an automated mess.' I cannot sum it up better.

Education, Training, Research and Development

To transmit from one generation to the next the accumulated wisdom and knowledge of the society, and to prepare young people for their future membership of the society and their active participation in its maintenance and development.

President Julius Nyerere

Education is generally, and often somewhat uncritically, considered to be a 'good thing'. It is important that one should question the accuracy and value of some of these 'universal truths'. For example, raw statistics on numbers of children in school do not give any indication of the quality, standard, content* and relevance to the students' future careers. Man-power planning is a notoriously inaccurate and difficult exercise but some attempt should be made to assess future needs and relate them to the educational system. The unemployed school leaver is a world-wide problem. Tett[1] quotes the Kenya development plan of 1970–74 where it is stated that 'of the 138,000 secondary school leavers, 70,000 of whom will have completed four years' secondary education, perhaps less than 50 per cent will obtain wage employment'.

Higher education too must be related to job opportunities. Tett[1] quotes the fourth five-year-plan of the Indian Government as saying 'During the course of the third five-year-plan there was a considerable expansion in facilities for engineering education at diploma and degree level. In the year of publication of this plan (1969) 45,700 engineers with degrees and diplomas were put out into a labour market already containing 20,000 unemployed engineers.' In planning educational systems it is

---

* One overseas educationalist told me that he discovered a small village primary school in the Far East where the children were reciting the names and dates of the kings and queens of England.

essential to consider relevance, needs, job opportunities, and not least the provision of suitably trained teachers and the associated buildings and equipment.

The resources available for education in the developing countries are relatively meagre and this situation is unlikely to change for some considerable time. The situation is likely to be further worsened by the expected population growth. UNESCO (United Nations Educational, Scientific and Cultural Organisation)[2] estimates that the population aged between 5 and 14 years in the developing countries will increase in the next decade from 550 million to 725 million. About 40 per cent of these children may be expected to attend school. It is therefore of great importance that the funds which are available for education are used to maximum effect. This requires that a number of difficult policy decisions be made, including the allocation of resources between the primary, secondary and higher educational sectors, and also on the relative weighting between general and vocational education. The relevance of the syllabus to the needs of the children and the efficiency of the teaching processes are also of concern. Fewer than half the children in the poorer countries attend school and of these only a small fraction move on to secondary level. The drop–out and repeat levels are also disturbingly high. Because of malnutrition and associated illness, the attendance is markedly less than in developed countries and performance and attention in class is also adversely affected. In education, as in other respects, the rural areas are in a far less satisfactory state than are the urban areas.

In setting up a primary syllabus it is necessary to recognise that this will be the only education received by most of the children, and the topics chosen should be the most relevant to the child in his activities on leaving school. Too much emphasis is often placed on preparation for transfer to secondary school. Consideration of the very high percentage of children who receive no formal education at all suggests the need for an alternative educational package that is more universally available. The roles of both traditional and adult education should not be neglected either when setting up an integrated educational system. The statistical evidence on which these remarks are based is given in appendix VI.

In this chapter we look at some of these matters and endeavour to identify areas in which Appropriate Technology techniques may be of assistance.

Traditional versus Western education

As Tett points out, all human societies have evolved systems for the transmission of skills, wisdom and way of life. The role and value of traditional forms of education are often overlooked in the consideration of educational systems. Traditional education may be divided into three phases:

(1)   Family.
(2)   Informal education from the community.
(3)   Formal development rites.

Traditional education teaches social customs, religious beliefs, tribal laws and civic customs, skills such as building houses, rearing the family, tending domestic animals and agricultural practices. Technical skills such as blacksmithing are usually handed down from father to son as a family tradition. This unstructured but effective method of education represents a considerable educational resource—not least, it prepares people for life in their own community.

Educational practices, particularly in the ex-colonial countries, have been based on the Western pattern and are aimed at producing the potential white-collar worker. Much of this training is unsuited to meeting the problems of both urban and rural employment. The two systems are not necessarily incompatible; both have a part to play.

Informal education

As already pointed out, only 40 per cent or so of children in the poorer countries actually receive any formal education. In addition, there is the problem of the unemployed school leaver and also the need to train adults in new practices and techniques.

A recent study prepared for United Nations Children's Fund (UNICEF)[3] has defined what it calls the 'minimum learning needs' to enable an individual to participate effectively in the economic, social and political activities of the community. The 'minimum learning' package includes 'functional literacy and numeracy (skill in using numbers), knowledge and skills for productive activity, family planning and health, child care, nutrition, sanitation and knowledge required for civic participation'. This basic education is not, of course, confined to children of normal school age but is also applicable to adult groups.

There are three principal groups of persons who benefit from informal education:

(1) Those who have not attended school.
(2) The unemployed school leaver.
(3) Adults requiring new skills.

There are currently many exciting experiments in informal education taking place, three of which are described below.

Futuro para la Ninez (Futuro)

Futuro para la Ninez—'future for the children'—is a private Colombian organisation founded by a group of people who believe that children are capable of awakening in adults a strong desire to act immediately to improve the conditions in which their children live.

The organisation, which has now been operating for several years, was founded by the sociologist and educator, Dr Richard Saunders. Futuro does not make gifts or awards since these are felt to lessen the initiative and desire to work of a community. Instead the organisation's officers try to create conditions which will increase the creativity and working capacity of the community.

The group is made up of highly selected community counsellors who have been specially trained by Futuro. The counsellors include sociologists, agronomists, peasants, educators, engineers and administrators. The policy of Futuro is to tap the practical intelligence of the peasants and get them involved from the very beginning in activities which will benefit their children and their community. The method of working adopted by the counsellors is unusual. They visit communities where they gather together groups of peasants and ask them the basic question: 'Are you satisfied with the conditions in which your children live? If not, what would you like to do to improve the situation?'

The peasants, with their local knowledge, can establish priorities which may be more meaningful than those established by visiting sociologists and technologists. Having decided on a need—for example, a water supply—the peasants meet Futuro counsellors to plan the work. The counsellors never provide solutions, but ask questions which stimulate and guide discus-

sion by the peasants who eventually arrive at a solution themselves. In this way they get used to thinking independently and
are willing to work on something that has been conceived by
themselves. This method builds up self confidence, enhances
human values and changes social attitudes.

Projects carried out by Futuro include school building (figure
9.1), fish farming, dry latrines, and terracing to prevent soil
erosion (see chapter 4). Health is of great concern to the peasants
in Colombia; over 80,000 children of less than 5 years die each
year, and most peasant families have lost at least one or two
young children. Futuro have encouraged the setting up of health
care in a similar manner to the 'barefoot doctor' scheme described in chapter 7.

The Futuro work is a unique and outstanding development
and could well be extended to other parts of the world.

FIG. 9.1   Futuro village school *(Photo Futuro)*

Village polytechnics

One solution to the problem of unemployed school leavers is the
'village polytechnic'; in Kenya the Ministry of Cooperatives and
Social Services is promoting the village polytechnic concept put
forward by the National Christian Council of Kenya. Rural
youth needs training in skills other than purely agricultural
activities, and two types of organisation have been set up.

(1) Institutional organisations—Here students are taught crafts such as carpentry, building, weaving and horticulture.

(2) Extension—Projects are initiated on the students' family land and are concerned with cash crops, bee-keeping, poultry, construction of fish ponds, etc.

The experience to date has been encouraging (figures 9.2, 9.3, 9.4). The training methods are of low cost compared to conventional training and the material taught is relevant to the needs of the student. Similar initiatives have been taken elsewhere in both rural and urban areas; as an example of the latter the work of the S.E.N.A. organisation in Bogota, Colombia, should be mentioned.

FIG. 9.2   Industrial training centre (Mombasa, Kenya) *(Photo Church Missionary Society)*

Rockefeller Rural Development University

This unusual experiment was initiated by the director of the Rockefeller Foundation in Cali, Colombia, Dr Arbab. Dr Arbab, who is by training a physicist, founded his requirements in the training of engineers intending to work in rural areas on two premises.

FIG. 9.3   Bicycle repair class (Naro Moru, Kenya) *(Photo Church Missionary Society)*

FIG.9.4   Rabbit rearing (Naro Moru, Kenya) *(Photo Church Missionary Society)*

(1) Such engineers, of university degree level, should be trained in the type of rural area in which they were intending to work.

(2) Young people brought up in the country were more likely to remain there if they were never removed from their rural surroundings.

An area of some 90,000 people was selected as the site for the new institution and very simple single-storey buildings erected using local building methods and materials. The buildings included both teaching and living accommodation. The students, 27 in the first year, came from local villages and had initial qualifications somewhat lower than those required for normal university entrance. The syllabus coverage is remarkably wide and includes agricultural botany, agriculture, rural health, sanitation, low-cost building and other topics. The permanent staff of six comprises one anthropologist, one physicist/biologist, one mathematician, two mechanical engineers and one agricultural engineer. Further assistance in teaching is provided by staff from the nearby university. The course extends over six years: the first two years bring the students to normal university entrance standard; the second two years consist of courses selected from the undergraduate courses at the university but chosen for rural relevance; and the final two years are of a problem-solving practical nature. An example of the latter might be the solution to a local village water and sanitation supply problem. It is of course too early to be able to comment on this interesting and ambitious scheme, but initial progress has been encouraging. One question which will need to be answered is whether this broad, rather superficial treatment can be taught in a manner which includes the discipline and rigour of a normal university course.

Such initiatives are to be welcomed and could make a very important contribution to education and training in the developing countries; however, conventional teaching is the dominant teaching method in all countries and it is not possible to make sudden changes even if they were felt to be desirable. The use of audio and video tapes and film strips provides a powerful and effective method of introducing new material and approaches. At university level the Open University material could be very valuable.

We will now look at ways in which current teaching practice might be assisted.

Primary education

A major problem, particularly in rural areas, is the provision of teachers. It may be that a para-professional approach similar to that adopted for paramedical staff could be more widely adopted. However, unless such training is very carefully thought out, the overall effect could be negative in value. School buildings need not be expensive and the cost can be reduced greatly by the adoption of low-cost building techniques as described in chapter 7.

A general criticism of education in developing countries is of the lack of training in practical skills.* All children in developed countries have direct experience of devices used in the home in the form of domestic equipment and also will have played with constructional toys—for example, meccano. One finds that quite small children presented with a nut on a bolt are able to unscrew it. The general public also make great use of the Do It Yourself market and examples are found in most homes.

This background experience is often lacking in developing countries and its educational contribution is undervalued. There is a possible role here for members of the university education department, in association with their engineering or physics colleagues, to design simple equipment. Such equipment could be manufactured cheaply by local craftsmen, and the work coordinated by a product development unit of the type described in chapter 8. The existence of the latter is not essential to the project but simply helpful if it happens to be there. Of course, the provision of equipment is not in itself sufficient and it is essential that the value of practical training be understood and accepted by the teacher. Such ideas should be introduced in the initial training courses, and also by 'in service' training.

Schools can also be used as another means of introducing a new technology or development into a community. One example is in the setting up of kitchen gardens to provide school meals but which allow the introduction of plants new to the area and also show the children how these are sown and tended. The Botswanan water-tank project described in chapter 1 not only introduced the water tank to the community but also provided the water supply for such a school kitchen garden. A general problem in many developing countries is in the provision of

---

* That there is no lack of inherent ability will be known to anyone who has watched a roadside car mechanic at work, for example in West Africa.

firewood for domestic cooking. Severe erosion problems exist because of indiscriminate chopping down of trees and other vegetation, and the disappearance of the trees necessitates long walks, often of several miles each day, to obtain fuel supplies. One alternative is the solar cooker. Such cookers could first be introduced in the schools for school meals and in some cases the simpler forms of cooker might actually be constructed in the school.

Secondary education

Most of the remarks made for primary education also apply to secondary education, in particular the provision of equipment. A local university can be of great assistance; for example, the Physics Department of the University of the West Indies, on the Jamaica Campus, has set up a Schools Science Unit which provides both equipment designs and also training and a technical back-up service to science teachers. Simple equipment suitable for school science laboratories include metre rules, retort stands, a bio-gas source for use with bunsen burners, and a solar still to provide distilled water.

It is at this educational stage that the question of vocational versus general education arises, and the job prospects for the students and the relevance of the syllabus to these jobs must be carefully considered.

Technical colleges or universities?

The shortage of a labour force trained to the level of craftsmen and technicians is a serious problem in most developing countries, and is exacerbated by the academic and non-practical approach of many of the university trained engineers and technologists in supervisory positions. Not only is there a shortage in supply of trained craftsmen and technicians but, for reasons of both social prestige and money, a significant number are attracted to white-collar jobs on completion of their courses. One way to get more trained craftsmen is to encourage on-the-job training in industry, but unfortunately such opportunities are scarce in most of the developing countries. Whilst the university position is well catered for in the developing countries, there is a need for more emphasis to be placed on the provision of more craft and technician training institutions. The equipping of the workshops and other facilities of such institutions provides a good opportunity for A.T. solutions. One difficulty is the provision of suitably trained instructors. In some countries the armed forces could perhaps provide such staff, and training could form part of compulsory military service; another initiative has been taken by the International Labour Office who have established an excellent training centre for technician instructors in Milan.

The universities — vocational or general?

The university has a number of functions, not least of which is the study and recording of local culture, traditions and values. The wisdom of setting up in developing countries departments in some of the currently fashionable or esoteric disciplines in developed countries is questionable. The imbalance between numbers of students taking law, sociology and the humanities compared to those taking engineering, science, agriculture and medicine is a cause for serious concern. It is important to bear in mind the cost of keeping a student at university in terms of the number of farmers necessary to support him. The cost per student of higher education is many times that for primary education.

The training of engineers and technologists

One major function of a university is the production of suitably trained undergraduates. The courses may be of a specialised nature or may provide a more general background training; courses may also differ in the degree of practical emphasis included in their teaching. In starting our new department at Reading,[4] our initial aim, in 1965, was to set up courses suitable for engineers intending to enter the high technology fields such as nuclear engineering. The experience of some of us in these fields suggested that there existed a requirement for engineers with a good understanding of physical principles, and the ability to apply them over a very diverse range of engineering problems, rather than a need for narrow specialisation. Our syllabuses were therefore arranged to include basic physical principles and their application to a variety of engineering problems.

During their courses the students are made aware of the almost explosive rate of development of ideas, techniques and materials, so that when they are professionally employed they will continually update themselves. Students are also introduced to the importance of non-technical considerations — such as economic, environmental, social and other factors — which need to be taken into account in selecting a design solution. Creative design and practical skills are believed to be of great importance and are taught by means of practical projects. These projects usually are carried out by small groups of students, ranging in number from three to seven, and each project extends over one academic year. Some typical recent examples of projects include investigation into and assessment of:

> Water pumps for low-head/low-power applications.
> Human energy output.
> Vertical-axis water turbines for developing countries.
> The Humphrey pump for boat propulsion.
> Savonius rotor windmills.
> Vertical-axis windmills.
> Novel steam engine.
> Solar powered pump.
> Solar water heaters.
> Solar cooker using a heat pipe.

Appropriate Technology provides many useful under-graduate projects which have a number of desirable features:

(1)   The requirements are new to the student and solutions to the problems cannot be found in the standard text-books; thus there is scope for initiative and invention.

(2)   The projects are relatively small in scale and can therefore be completed in one year.

(3)   The results are of value.

Such projects can lead to work suitable for a higher degree and may serve to initiate a line of research for the member of staff supervising the project. (A list of project titles is given in appendix III.) We have found that the type of engineering course described above, though initially designed for engineers in the high technology sector, is in fact well suited to the needs of engineers intending to work in the developing countries. This is not particularly surprising, since such people are frequently presented with new or unfamiliar problems and lack custom-designed equipment and the availability of specialist help. Project teaching of the type described has been introduced, with success, in a number of universities in developing countries. One difficulty which can arise is the absence of practical experience of and familiarity with industry on the part of academic staff; the result can be a lack of confidence and a disinclination to move away from the security of standard textbook teaching. This is a very real problem to which there is no simple solution; however, the use of visiting lecturers to set up the initial projects, attendance by staff at workshops and symposia, and short attachments to organisations overseas can all be of value in overcoming this problem.

Many university departments suffer from a shortage of equipment; in others there is sometimes a tendency to buy expensive and often unsuitable equipment from overseas. The latter is understandable where lecturers are hard pressed and local design and manufacture facilities are not readily available. There is however much to be said for the encouragement of the use of equipment made in the engineering department's own workshops, and the setting up of such workshops should be given high priority. Buildings are often unnecessarily elaborate and expensive. For example, on a recent overseas visit I found a large and expensive building devoted solely to hydraulic experiments. Essentially this equipment consisted of a small stream of water into which was introduced a variety of dams, obstructions and constrictions. This arrangement does not need a building at all and could easily be situated in the open air (as indeed was such a laboratory in another university in the same country), and the building could then be used for apparatus which requires adequate weather protection.

The syllabus of a course should be carefully examined for local relevance (and irrelevance). Harry Dickinson once told me that, when lecturing to undergraduate courses in electrical engineering in tropical Africa, the syllabus he was required to teach included the effect of ice formation on transmission cables. This is quite an important factor in the design of such cables in temperate countries, but the likelihood of such a contingency arising in that area was, to say the least, somewhat remote.

Some universities in developing countries offer first degrees in Appropriate Technology. I have reservations about the desirability of such courses and suspect that they may lack coherence and depth. I would prefer to advocate a general degree of the type described above, in which the basic principles are taught and illustrated by examples of local interest, and where use is made of hardware-oriented (not paper study) Appropriate Technology projects. However, the inclusion of a course of lectures on Appropriate Technology as part of the General Studies section of many degree or diploma courses would be well worthwhile. A case can be made for a higher degree in Appropriate Technology, and a syllabus for such a degree has been prepared in my own department.

Research and development

Research and development is carried out in universities and research institutes. In the former there is often a need for much more relevance in the choice of topic, whereas the latter frequently do not pay sufficient attention to ensuring that the results of their work are taken up by industry.

Often a postgraduate, on returning to his university, attempts to continue with his original line of research and one finds research activities such as nuclear fine structure or magneto-hydrodynamics carried out in isolation from other workers in the subject, and in countries where there are much more pressing and immediate local needs. That a worker should wish to continue with a familiar line of activity is understandable and the fault lies in the choice of his original topic (an example of what I once heard described as promoting the very best of what is wrong with the Western educational system). This difficulty may be further compounded by the attitude of colleagues and the university authorities to more applied small-scale work; the latter may be regarded as substandard and unsuited for university research. It should be stated firmly that Appropriate Technology problems can present an intellectual challenge just as great as that presented by 'Big Science and Engineering'—the boundary conditions and scale only are different.

How does one start up a research programme in an Appropriate Technology subject in a university? It is certainly difficult: not least is the problem of finding colleagues interested in discussion of the programme. For the young man, one method is for him to carry out a programme of research on a suitable topic in another university, possibly to obtain a higher degree, and on return to his own university to continue the work while at the same time maintaining the links with his previous colleagues. Some universities allow higher-degree students to complete the work back in their home university, where good

reason can be shown for this. In the case of Appropriate Technology there is considerable merit in this arrangement since the
home university can provide the appropriate environment for
the field testing of the equipment. The setting up of formal or
informal links between different universities is a very valuable
mechanism for promoting A.T. type research, and many of us
are involved with such links and receive great benefit from them.
University staffs are usually fairly heavily committed to their
teaching, research and administrative duties, and do not have a
great deal of spare time for other activities. On the other hand, in
developing countries they often represent the major source of
technical and scientific expertise. The product development
centre concept discussed in chapter 8 provides a mechanism for
making use of this expertise in an effective manner and with the
minimum loading on the individual members of university staff.

The other source of development potential is the research
institute. Some of these are exceptionally useful and
relevant—for example, the I.A.T. Centre in Ibadan, Nigeria.
Unhappily, this is not always the case. The principal criticism I
would make of some research institutes is not on the quality of
their work, which is often very good, nor of their choice of
research topic, but on how their developments are used—or not
used. Figure 9.5 shows what often happens: a need is identified
and a solution is suggested. A programme of work is initiated,
culminating in the development of a prototype machine, which
is then tested, modified and eventually pronounced satisfactory.
What follows is a paper to a conference and the device is then
placed in the establishment's museum. This 'museum filling' is
quite useless. It is essential that from the beginning of a project
attention must be paid to the potential users, who must be

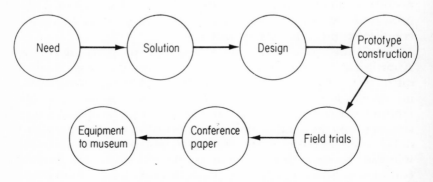

FIG. 9.5   A typical development flow chart

introduced to the development, and also to local manufacture possibilities so that a market can be built up together with a suitable source of supply. The same criticism can also be made of the results of much university research. This question of the take-up by industry is of fundamental importance and is too often completely overlooked.

Getting Started

I want to change things. I want to see things happen. I don't want just to talk about them.

J. K. Galbraith

There is a need for more information to be made available to the general public on the subject of Appropriate Technology, in both the developed and the developing countries. In the developing countries a better understanding and acceptance of the Appropriate Technology methods is required; this is a problem of general education. The journal *Appropriate Technology*, published by I.T.D.G., is a good source of information on A.T. developments. In order that Appropriate Technology should develop, it is also necessary to have a favourable economic and political climate, which will take different forms and priorities in different areas.

Appropriate Technology is not necessarily confined to small-scale solutions, and there are circumstances in which a power station or a large dam may be the most appropriate solution. Such projects are well understood, technical experts are available and methods of funding and implementation well known. In this book we are concerned primarily with developments in rural areas where small-scale, self-help techniques are usually more appropriate. Small manufacturing concerns may prove more economic than larger central factories, in some situations, particularly when full social costs are taken into account. In some of the published comparisons between large and small industry the former appears to be more economic because full production has been assumed and allowance has not been made for such factors as breakdown and distribution costs. Examples of small industries which have proved competitive include brickmaking and cement manufacture.

Information and advice on specific problems may be obtained from both regional and international organisations. Examples of the former include the Technological Consultancy Centre in Kumasi, Ghana, and the Appropriate Technology Development Association, Lucknow, India. International organisations include I.T.D.G. in London and the Brace Institute in Canada. Addresses of a number of regional and international bodies are given in appendix VIII. Some of these organisations, such as I.T.D.G. carry out field projects and can arrange for visits by consultants and other staff.

Before approaching an organisation with a query, anyone who is interested should carry out background reading in the subject. General references are given at the end of this book. He should also give information on local conditions and requirements. In I.T.D.G. we often receive letters asking for 'full information on methane generators, solar heaters and windmills'. Such letters are easy to write but not to respond to—at least not politely. We can, sometimes, provide details of particular designs which may be suitable. In some cases the designs can be employed without significant modification, in others they require to be altered to suit local materials and manufacture capabilities. There are no universal solutions.

Schools, colleges, universities and other educational establishments in developing countries

A.T. concepts can be introduced into the curriculum and local needs provide subjects for useful projects at all levels. Staff can form interdisciplinary groups. University and other research groups can select A.T. topics for their research programmes, and as has been said earlier, these prove most suitable for postgraduate studies. The setting up of links with university departments with similar interests and situated in the developed countries can be of great assistance in establishing a new research school. Some universities may wish to set up product development centres of the type described in chapter 8.

Experts in developed countries

For those who work in a university or similar establishment, or who have relevant facilities or expertise, their help will be very much welcomed in the work of groups such as I.T.D.G. Newcomers should bear in mind that it is essential that any design recommended for use in developing countries should first be rigorously tested, both locally and overseas; this has not always been the case in the past.

I hesitated over the word 'expert' since it is often (rightly) used in a pejorative sense—'An expert is a man who has stopped thinking—he knows' (F. Lloyd-Wright). However, I chose it deliberately because groups such as I.T.D.G. are under-staffed and hard pressed, and we can only usefully employ volunteers who have relevant expertise and also are prepared to devote a significant amount of time to our work.

Voluntary work overseas

Younger folk may wish to work overseas for a period, which can be both useful to the recipients and a valuable experience for the individual. There are several national organisations which make such arrangements, such as V.S.O. (Voluntary Service Overseas) in the U.K. and S.V.S. in the U.S.A. Volunteers should be quite clear about their motivation and role; they should want to help and be prepared to work at a level appropriate to their skills and experience. Developing countries have quite enough problems of their own without having to sort out our misfits. Attitudes should be modest in accord with ability.

Some years ago I sat on a Round Table discussion as part of a briefing meeting for volunteers for overseas; I was expecting to be questioned on such topics as the treatment of the minor ailments of the diesel engine. Instead the first question addressed to me was on the 'ethics of the profit motive in developing countries', and the second question was 'what part should I play in the political development of the country?' The answer to the second question is, of course, 'None, unless you wish your stay to be very short or unduly prolonged'. Fortunately this attitude to the job is not typical. Volunteers should take a broad general interest in the country and its problems, but must not forget that they are there to do something useful in a short time.

Private fund raising

National and international aid is summarised in appendix VII. When faced with the enormous sums which governments, United Nations bodies and other organisations allocate to aid developing countries, the private fund raiser may well question whether the small amounts he can gather are really worth the effort. They are. There is a rather curious relationship between the total sum required to support a project and the difficulty in achieving it which I call the Law of Inverse Difficulty: £1,000,000 is relatively easy to get for a programme, £100,000 is rather difficult, £10,000 is very hard indeed, and £1,000 is virtually impossible. This is because the large aid organisations like to have a few prestigious projects of large size which are relatively easy to control financially. On the other hand those of us working in Appropriate Technology may wish to send a windmill head costing a few hundred pounds to the Sudan, or a Humphrey pump to Bangladesh, or an adviser, often at short notice, to Nepal. It is here that the voluntary organisations are often so helpful. Keep it up.

Postscipt

Chesterton once wrote 'If a thing is worth doing, it is worth doing badly' with the implication 'rather than not at all'. This is a most penetrating remark and is not a bad philosophy of life. It might perhaps be rephrased a little more positively: in Appropriate Technology we try to make the best use of the resources available to us and throw off the constraints of conventional solutions.

Appendix I

Conversion Factors

The International System of units, or SI system, has been used in this book (it is also called the metric system). Although this system is now becoming increasingly adopted the older Imperial units are still extensively used. People accustomed to thinking in one set of units have difficulty in visualising quantities which are expressed in a less familiar system.

In this appendix, conversion factors are given to enable Imperial measures to be converted to the SI System and vice-versa.

Quantity	SI unit and symbol	Imperial units to SI units	SI units to Imperial units
Length	metre m centimetre cm millimetre mm kilometre km 1,000 m=1 km 1,000 mm=1 m 100 cm=1 m	1 inch=2.54 cm 1 ft=30.5 cm 1 yd=0.914 m 1 mile=1.61 km	1 cm=0.394 in 1 m=3.28 ft 1 km=0.62 miles
Mass	kilogram kg gram g tonne t 1,000 g=1 kg 1,000 kg=1 t	1 lb=454 g 1 ton=1.02 tonne	1 kg=2.2 lb 1 t=0.98 ton
Area	square metre m^2 square kilometre km^2 hectare ha 1 ha=10,000 m^2 1 km^2=1,000,000 m^2 =100 ha	1 ft^2=929 cm^2 1 yd^2=0.836 m^2 1 acre=0.405 ha 1 $mile^2$=2.59 km^2	1 m^2=10.8 ft^2 1 ha=2.47 acres 1 km^2=0.386 $mile^2$

Quantity	SI unit and symbol	Imperial units to SI units	SI units to Imperial units
Volume	cubic metre m³ cubic centimetre cc litre 1,000 l=1 m³ 1,000 cc=1 litre	1 pint=0.568 l 1 Imp. gal=4.55 l 1 cubic foot=28.3 l	1 l=1.76 pints Imp. 1 l=0.22 Imp. gal 1 l=0.26 U.S. gal
Flow Rate	litre per sec l/s cubic metre per sec m³/s 1 m³/s=1,000 l/s	1 gal/min=0.076 l/s	1 l/s=13.2 gal/min
Energy	joule J megajoule MJ gigajoule GJ 1 MJ=1 million J 1 GJ=1,000 million J 1 kWhr=3,600,000 J	1 B.T.U.=1,055 J	1,000 J=0.95 B.T.U.
Power	watt W kilowatt kW 1 kW=1,000 W	1 h.p.=0.746 kW	1 kW=1.34 h.p.

Appendix II

Gross Domestic Product, Gross National Product and National Income

Gross Domestic Product (G.D.P.)

The G.D.P. of a country is the total value of all goods and services made available for consumption or for adding to wealth in one year.

Gross National Product (G.N.P.)

The G.N.P. of a country is the total national income of that country in one year. It differs from G.D.P. in that it includes returns from investment overseas but does not include earnings by foreign investors or non-nationals within the country.

G.N.P. = G.D.P. + investment returns from overseas − returns from foreign investment in the country

The G.N.P. includes the incomes of all residents, companies and government bodies. The value of G.N.P. divided by the population gives the G.N.P. per capita for the country and is usually regarded as an indication of the national standard of living. The G.N.P. per capita is the average income for each man, woman and child in the country and should not be confused with average wage.

Figures for G.N.P. per capita should be treated with some caution for the following reasons

(1) It is difficult to obtain accurate statistics in many of the developing countries, and the non-market component, which is estimated, may make up a large fraction of the total.

(2) The figures are average values and may conceal large variations.

(3)   The figures are converted to a standard currency, usually dollars, and the official rate of exchange may be set at an unrealistic value.

However, whilst noting these defects, the G.N.P. per capita remains a useful indicator of general standard of living for purposes of comparison.

National Income

Some authors use National Income and Gross National Product interchangeably, others use the term National Income to mean G.N.P. less depreciation on capital goods. Samuelson[1] calls this 'subtracting the deathrate of equipment'. The 'birth rate of equipment' is already included in the value for G.N.P. Apart from figure 2.1 which uses National Income per capita as representing G.N.P. less capital depreciation, all other figures in the book are given in terms of G.N.P.

Appendix III

Project Proposals by I.T.D.G. University Liaison Unit[1]

AGRICULTURAL
    Animal drawn harvester
    Mouldboard ploughs
    Cotton seed drill
    Green manure trampler
    Baby tractors
    Portable corn grinding machinery
    Sugar cane crusher

POWER
    Paraffin or butane refrigerators
    Capabilities of draught animals, training and equipment
    Windmill operated generator
    Pedal operated electric generator
    Small portable steam engine using low grade fuel
    Solar cookers
    Methane gas from manure etc.

CHEMICAL
    Hand operated plastics machinery
    Distilling or refining spirits
    Vegetable oil processing
    Small petro-chemical project based on refinery by-products
    Refining used lubricating oils
    Reactive dye equipment and materials
    Purifying sheep lanolin for shampoo or cream

FOOD INDUSTRIES
    Oil expelling machinery for oilseed crops
    Canning machinery for fruits and juices
    Vegetable processing unit
    Tomato canning machinery
    Groundnut roasting machinery
    Bakery equipment
    Manufacture of glucose from tapioca starch

BUILDING, WOOD AND PACKING
Saw milling and woodwork
Radio cabinets
Building fittings
Chip particle and sawdust boards
Fibre board from groundnut husks
Hessian sack manufacturing
Corrugated cardboard packing box machinery

TEXTILES
Garment making
Tent making
Cotton ginning and pressing
Cordage, ropes, twines etc.
Surgical cotton
Hosiery
Waste cotton utilisation

POULTRY
Incubators
Sterilisers
Bird crates
Chick sexing machines
Poultry weighing scales
Automatic chain feeders

OTHERS
Zip fastener machinery
Electroplating equipment
Washing machine
Iron smelting
Reconditioning of car batteries
Book matches from waste paper
Soap manufacture

Appendix IV

Water-related diseases

The Bradley classification of water-associated diseases was given in chapter 5. Table 1 (taken from reference 1) lists these four mechanisms together with suitable preventative measures. Since some infections can also be transmitted by faeces the faecal-oral diseases should be added to both the water-borne and to the water-washed categories. Feachem has proposed the classification of table 2 to overcome this difficulty. Table 3 lists the main water-related diseases with their water associations and pathogenic agents.

Table 1[1]. The four mechanisms of water-related disease transmission and the preventive strategies appropriate to each mechanism

Transmission mechanism	Preventive strategy
Water-borne	Improve water quality Prevent casual use of other unimproved sources
Water-washed	Improve water quantity Improve water accessibility Improve hygiene
Water-based	Decrease need for water contact Control snail populations Improve quality
Water-related insect vector	Improve surface water management Destroy breeding sites of insects Decrease need to visit breeding sites

Table 2¹. A classification of water-related diseases

Category	Example
1. Faecal–oral (water-borne or water-washed	
(a) low infective dose	Cholera
(b) high infective dose	Bacillary dysentery
2. Water-washed	
(a) skin and eye infections	Trachoma, scabies
(b) other	Louse-borne fever
3. Water-based	
(a) penetrating skin	Schistosomiasis
(b) ingested	Guinea worm
4. Water-related insect vectors	
(a) biting near water	Sleeping sickness
(b) breeding in water	Malaria

Some common water-associated diseases and their characteristics

Water-borne or water-washed

Diarrhoeal diseases are caused variously by bacteria, viruses and protozoa, and together make up the most common group of water-associated diseases in tropical countries. There can be few of us who have not at some time experienced a mild form of this complaint; those who have travelled in tropical countries will probably have unpleasant memories of some of the more severe forms, known under various guises—Montezuma's Revenge, Hong-Kong Dog, Aztec Two-Step (wonderfully descriptive). Diarrhoea, dysentery, gastroenteritis are debilitating complaints which occur widely over the developing world.

In addition to the endemic water-borne diseases mentioned above, there are also serious epidemic diseases such as typhoid and cholera which are associated with contaminated water and can cause high local death rates.

Water-washed diseases

This category might better be described as water-unwashed diseases. Lack of water for purposes of hygiene leads to the transmission of eye infections such as trachoma which may

Table 3. Water–related diseases with their water associations and their pathogenic agents[1]

Water–related disease	Category from Table 2	Pathogenic agent
Amoebic dysentery	1b	C
Ascariasis	1b	D
Bacillary dysentery	1b	A
Balantidiasis	1b	C
Cholera	1a	A
Diarrhoeal disease	1b	H
Enteroviruses (some)	1b	B
Gastroenteritis	1b	H
Giardiasis	1b	C
Hepatitis (infectious)	1b	B
Leptospirosis	1a	E
Paratyphoid	1b	A
Tularaemia	1b	A
Typhoid	1a	A
Conjunctivitis	2a	H
Leprosy	2a	A
Louse–borne relapsing fevers	2b	E
Scabies	2a	H
Skin sepsis and ulcers	2a	H
Tinea	2a	F
Trachoma	2a	B
Flea-, louse-, tick- and mite-borne typhus	2b	G
Yaws	2a	E
Clonorchiasis	3b	D
Diphyllobothriasis	3b	D
Fasciolopsiasis	3b	D
Guinea worm	3b	D
Paragonimiasis	3b	D
Schistosomiasis	3a	D
Arboviral infections (some)	4b	B
Dengue	4b	B
Filariasis	4b	D
Malaria	4b	C
Onchocerciasis	4b	D
Trypanosomiasis	4a	C
Yellow fever	4b	B

A=bacteria; B=virus; C=protozoa; D=helminth; E=spirochaete; F=fungus; G=rickettsiae; H=miscellaneous.

produce partial or even complete blindness. Skin complaints are also widespread where adequate water for washing is not available.

Water-based diseases

These are all due to worm infections. One of these, *Schistosoma*, makes use of a snail host and has a somewhat complicated life cycle; the guinea worm, another common infection, requires a small crustacean as its host.

Schistosomiasis or bilharzia is a disease caused by a parasitic worm or blood fluke, which enters the human through the skin from infected water. The life cycle of the fluke is shown in figure IV. 1. The intermediate (snail) host lives in the relatively still waters of canals, ponds, slow-moving rivers and backwaters.

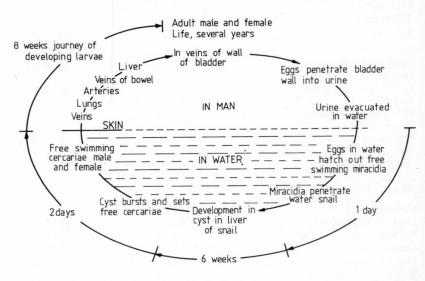

FIG. IV.1   The life cycle of Schistosoma haematobium (Bilharzia). A blood fluke

The symptoms of schistosomiasis include enlarged spleen or liver, dropsy, cystitis and progressive anaemia. The disease occurs widely in Africa, the Middle and Far East and parts of South America. The life cycle can be broken by eliminating the snail host by the use of molluscides. An interesting observation[2] reported from East Africa is that the snails were destroyed when local women used berries from a 'soap plant' to do their washing in the river. There are a number of plants around the world which contain such 'saponins' and it is possible that these could be developed and used to supply sufficient molluscide to destroy the snails. The investigation of this possibility provides a good example of research appropriate to a university in a developing country.

With guinea worm (dracunculiasis), the adult worm lies below the skin of its victim usually on the lower leg. When water is spilled on the leg guinea worm larvae may be released, and if this occurs by a well the larvae will be washed into the well. In order to survive the larvae require a host, a small crustacean known as cyclops, which acts in a similar manner to the snail host for Schistosoma. If people drink the water containing the infected cyclops the cycle is completed.

Water-related insect vectors

Malaria, a disease producing an acute fever, is transmitted by the bite of an infected mosquito. The larvae of the mosquito live in stagnant water. Filariasis (elephantiasis) is also spread by mosquito. The worms obstruct the lymphatic system causing fluid to accumulate in the legs and external genital organs, sometimes with the bizarre results suggested by its alternative name, elephantiasis.

Yellow fever and dengue fever are virus infections carried by mosquitoes whose larvae live in stagnant water often in domestic storage vessels.

Sleeping sickness is a serious infection transmitted by the tsetse fly which lives around stagnant pools and water holes.

Onchocerciasis (river blindness) is transmitted by a small black fly which breeds in fast flowing water. This is common in the savannah region of Africa and in its extreme form causes total blindness. In some villages up to a quarter of the population may be blind from this cause.

Diseases due to defective sanitation

These diseases include hookworm and roundworm. Hookworms exist in damp soil, for example around wells, and can penetrate the skin to enter the system. They cause major blood loss and can lead to anaemia. Roundworms can be transmitted by, for example, dirty food; the effect is to divert food from the victim.

Appendix V

Energy, Work and Power

Energy

The amount of energy we have is a measure of the work we are able to perform. We can regard energy as rather like the capital in our bank balance, and the energy equivalent of spending money is in doing work. The simplest example of performing work is that of raising a weight. In this case, if we raise the weight through a vertical distance by exerting an upward force equal to the weight, then the product of the weight and the distance will be the work done, or the energy required. This is all we need to know in order to estimate the energy required to pump water to a height.

In the SI or International System of units, energy and work are measured in joules (J). For example, the energy required to lift one cubic metre of water a vertical distance of 3 m will be given by the weight of the water multiplied by 3 m. One cubic metre of water has a mass of 1,000 kg and its weight is obtained by multiplying by the acceleration due to gravity (g), that is, $1,000 \times 9.81$ or 10,000 approximately. Hence the work required to lift the water will be $3 \times 10,000$ or 30,000 J which will be equal to the energy used.

Power

The rate at which we spend our money is given by dividing the amount of money spent by the time required to spend it; similarly the rate at which we spend energy is given by the amount of energy used divided by the time in which it is spent; we call this power. Power, which is joules divided by seconds, is measured in watts in the SI system. The kilowatt or kW is also used (1 kW is equal to 1,000 W). Another common unit of power is the horse power, which was first introduced by James Watt,

190

the inventor of one of the early steam engines. One horse power is very nearly three–quarters of a kilowatt.

Returning to our calculation on pumping, the power required if we wish to pump this amount of water every hour is given by 30,000 J divided by 3,600 seconds. (It is most important always to use the same set of units; in the SI system the unit of time is the second.) Hence the power works out as $8\frac{1}{3}$ watts, or slightly more than 0.1 of a horse power. A man working perfectly efficiently should be able to work at about half this rate for an eight hour day. In practice the equipment he will use will not be very efficient, and he may only achieve a quarter of this value or even less.

Conversion of energy to heat

There is one further important aspect of energy worth mentioning, and that is the conversion of one form of energy to another. In the previous example we have performed work, using an equal amount of energy, to raise a quantity of water. If this water has been stored in a tank it is possible (ignoring losses through friction) to get back all the original energy, for example by allowing the water to flow back to its original level, driving a water wheel on the way. In monetary terms this is merely like buying goods and then reselling them; one will not get such a good resale value, of course, and this resale loss is equivalent to friction in the water pumping case.* There is one important exception to this free conversion of one energy form to another — that is in the case of heat energy. Heat energy is rather like a soft currency: the exchange rate in the market will be much less than the official rate. In energy terms temperature is a measure of 'hardness' and a high temperature energy source, say 1,000°C, will convert quite well to other energy currencies, whereas a low temperature source, say 30°C, is practically worthless.

---

*Energy never appreciates; the quantity of energy never increases with storage, and in practice some may well be lost by various processes. Ideally the work done is equal to the energy consumed. This is known as the First Law of Thermodynamics or the Law of Conservation of Energy.

Appendix VI

Some educational statistics

Table 1 shows how enrolment ratio has varied over the decade 1960 to 1970 for the primary, secondary and higher educational sectors and for five groups of countries. The countries are grouped in terms of G.N.P. per capita, and enrolment ratio is the number of children in the appropriate age group actually at school divided by the total number in the group. There has been a general increase in enrolment but the level remains disappointingly low. For example, in the lowest income group, primary school ratios have increased from 34 per cent to 43 per cent, secondary from 4 per cent to 5 per cent, and higher from 0.3 per cent to 0.4 per cent. In secondary and higher education the gap between the highest and lowest income groups has increased.

Table 2 shows school textbook production for primary and secondary schools; we note that the eight countries listed as having a G.N.P. of under $250 per capita have an annual production of less than one textbook per student and some very much less than this figure. Also, textbooks are not necessarily relevant to the needs of the student. These figures reflect the amount of money available per student, and table 3 shows how this differs over the income groups. We see that there is a factor of roughly forty between group I and group V. Literacy is a good measure of the success of primary education; illiteracy in the over-15 age group decreased from 59 per cent to 50 per cent in the developing countries in the decade 1960 to 1970. However, due to population increase, the number of illiterates actually rose from 701 million to 756 million. The level of literacy in Africa is still only 25 per cent (see figure 2.6).

In chapter 9 it was suggested that educational facilities in the rural areas of developing countries are inferior to those in the urban areas. Table 4 compares the percentages of complete schools in these regions, rural and urban areas, broken down

192

under income groups and also geographic regions. A complete school is one which offers the full number of grades. Table 5 gives figures for the Latin American countries; in this table the numbers are given of those successfully completing a course, expressed as a percentage of the initial enrolment, and this table also gives the number of years required on average to complete the course successfully. The advantages of a student in an urban area over one from a rural area are clearly brought out by these figures. A very comprehensive list of educational data, taken from reference 1, is given in table 6.

Table 1 School enrolment ratios (from reference 1)

Per capita[1] G.N.P.	Number of countries	Total population in 1970 (millions)	Enrolment ratios[2]								
			First level			Second level			Third level		
			1960	1965	1970	1960	1965	1970	1960	1965	1970
I—Up to $120 (excluding India, Indonesia, Pakistan, Bangladesh)	25	168	34	39	43 (31)	4	5	5	0.3	0.3	0.4
India, Indonesia, Pakistan, Bangladesh	4	802	43	56	71 (63)	9	11	18	1.7	2.6	4.3
II—$121–250	23	287	67	79	83 (68)	9	14	19	2.1	3.0	5.6
III—$251–750	38	433	73	83	97 (77)	11	17	25	1.9	3.3	5.3
IV—$751–1,500	9	112	90	93	97 (80)	33	44	49	6.2	8.4	10.5
V—Over $1,500	24	623	100	100	100	58	65	83	17.0	23.7	30.2

[1] Countries in each group are as follows:

I—Afghanistan, Bangladesh, Botswana, Burma, Burundi, Chad, Dahomey, Ethiopia, The Gambia, Guinea, Haiti, India, Indonesia, Lesotho, Malawi, Mali, Nepal, Niger, Nigeria, Pakistan, Rwanda, Somalia, Sri Lanka, Sudan, Tanzania, Upper Volta, Yemen Arab Republic, People's Democratic Republic of Yemen, Zaire.

II—Bolivia, Central African Republic, Cameroon, Equatorial Guinea, Egypt, Ghana, Kenya, Khmer Republic, Republic of Korea, Liberia, Malagasy Republic, Mauritania, Mauritius, Morocco, Philippines, Senegal, Sierra Leone, Swaziland, Thailand, Togo, Tunisia, Uganda, Republic of Vietnam.

III—Algeria, Bahrain, Brazil, Republic of China, People's Republic of Congo, Colombia, Costa Rica, Dominican Republic, Ecuador, El Salvador, Fiji, Gabon, Guatemala, Guyana, Honduras, Iran, Iraq, Ivory Coast, Jamaica, Jordan, Lebanon, Malaysia, Mexico, Nicaragua, Oman, Panama, Papua New Guinea, Paraguay, Peru, Portugal, Romania, Saudi Arabia, Syria, Turkey, Uruguay, Yugoslavia, Zambia.

IV—Argentina, Chile, Cyprus, Greece, Singapore, South Africa, Spain, Trinidad and Tobago, Venezuela.

V—Australia, Austria, Belgium, Canada, Denmark, Finland, France, Federal Republic of Germany, Iceland, Ireland, Israel, Italy, Japan, Kuwait, Libya, Luxembourg, Netherlands, New Zealand, Norway, Qatar, Sweden, United Arab Emirates, United Kingdom, United States.

[2] The enrolment ratios have been obtained by dividing the total enrolment at each level with the appropriate age group. These 'gross' enrolment ratios are inflated by over-age students. For 1970, it has been possible to exclude the over-age students and estimate 'net' enrolment ratios at the first level. The net ratios are indicated in parentheses and show that the over-age students form 10–20% of the total student body at the first level.

Appropriate Technology

Table 2   School textbook production[1] (from reference 1)

Countries by per capita G.N.P.	Number of books (000's)	Enrolment (000's)	Textbooks per student
Up to $250			
Ghana	19	1,518	0.01
Cameroon	30	956	0.03
Nigeria	340	3,871	0.08
Uganda	259	768	0.34
Kenya	592	1,404	0.42
Tunisia	1,580	1,070	1.48
Sri Lanka	4,229	2,653	1.59
Egypt	9,694	5,187	1.87
			Average 0.73
$251–1,500			
Chile	1,695	2,345	0.72
Argentina	3,973	4,359	0.91
Malaysia	6,945	2,274	3.05
Singapore	2,396	513	4.67
Spain	30,592	5,879	5.20
			Average 2.91

[1] School textbooks for primary and secondary education.
Source: *Unesco Statistical Yearbook, 1972.*

Table 3   Public expenditure in education per student[1]
(from reference 1) (U.S. dollars current prices)

Countries grouped by per capita G.N.P.	1960	1965	1970	Net change
I Up to $120	16	21	18	+ 13%
II $121–250	33	40	49	+ 49%
III $251–750	43	58	57	+ 33%
IV $751–1,500	114	165	179	+ 57%
V Over $1,500	338	504	749	+121%
Group V amount as a multiple of Group I	21	24	42	

[1] Annex 9 contains additional data on education expenditures.

Table 4 Availability of primary schools in urban and rural areas
(from reference 1)
Percentage of the total number of primary schools in each category (urban and rural) which offer a complete number of grades

	Number of countries	Complete urban schools as a percentage of total urban schools	Complete rural schools as a percentage of total rural schools
Countries by G.N.P. per capita			
Up to $120 (excluding India)	9	53	36
India	1	57	49
$121–250	7	72	32
$251–750	16	77	62
$751–1,500	2	89	56
Over $1,500	6	100	99
By major regions			
Africa	16	79	54
Asia (excluding India)	9	94	66
India	1	57	49
Latin America	10	88	34
Europe	5	98	99

Source: Based on data in *Unesco Statistical Yearbook, 1972.*

Table 5 Comparison of education efficiencies in urban and rural areas in Latin America (from reference 1)
(a) Successful completers and dropouts in primary education

	Total country	Urban	Rural
	Successful completers	Successful completers as % of entrances	Successful completers
Colombia	27.3	47.3	3.7
Dominican Republic	30.4	48.1	13.9
Guatemala	25.4	49.6	3.5
Panama	62.3	80.7	45.3
Average percentage completers	39	51	22

(b) Efficiency of primary education

	Years taken to produce a successful completer				Input output ratio		
	Ideal	Total country	Rural	Urban	Total Country	Rural	Urban
Colombia	5	11	66	8	2.4	13.2	1.7
Dominican Republic	6	14	27	9	2.3	4.5	1.6
Guatemala	6	14	70	10	2.3	11.6	1.6
Panama	6	9	12	8	1.5	1.9	1.2

Source: Based on the Unesco report, *The Statistical Measurement of Educational Wastage.*

Table 6   Measures of health status by level of per capita Gross National Product (G.N.P.) in selected countries (from reference 1)

Country	Per[1] capita G.N.P.	Crude[2] birth rate	Crude[2] death rate	Infant[3] mortality	Life[2] expectancy
Burundi	60	41.8	24.9	150[4]	39.0
Upper Volta	70	48.5	24.9	180	39.0
Ethiopia	80	49.5	23.8	162	40.0
Indonesia	80	44.8	18.9	125[4]	45.4
Yemen Arab Republic	90	49.5	20.0	160	45.5
Malawi	90	47.7	23.7	148[4]	41.0
Guinea	90	46.6	22.8	240[5]	41.0
Sri Lanka	100	28.6	6.3	50	67.8
Dahomey	100	49.9	23.0	110[4]	41.0
Tanzania	110	50.1	23.4	122[5]★	44.5
India	110	41.1	16.3	139	49.2
Sudan	120	47.8	18.5	130	47.2
Yemen, People's Democratic Republic of	120	50.0	22.7	160	45.3
Uganda	130	46.9	15.7	160[4]	50.0
Pakistan	130	47.6	16.8	130	49.4
Nigeria	140	49.3	22.7	150–175[6]	41.0
Central African Republic	150	43.2	22.5	190	41.0
Mauritania	170	48.8	23.4	187[5]	41.0
Bolivia	190	43.7	18.0	60	46.7
Liberia	210	50.7	22.3	159	43.5
Sierra Leone	210	41.9	20.2	197[5]	43.5
Thailand	210	43.7	10.4	23	58.6
Egypt, Arab Republic of	220	37.8	15.0	120	50.7
Viet-Nam, Republic of	230	41.8	23.6	100	40.5
Philippines	240	43.6	10.5	62	58.4
Senegal	250	47.3	22.2	93[4]	42.0
Ghana	250	48.8	21.9	156[4]	43.5
Congo	270	45.1	20.8	180	43.5
Paraguay	280	42.2	8.6	39	61.5
Syrian Arab Republic	290	46.9	14.4	24[4]	53.8
Honduras	300	49.3	14.6	37[4]	53.5
Ecuador	310	41.8	9.5	87[6]	59.6
Tunisia	320	41.0	13.9	76	54.1
El Salvador	320	42.2	11.1	58	57.8
Ivory Coast	330	45.6	20.6	138[4]	43.5
Turkey	340	39.4	12.7	153	56.4
Algeria	360	49.4	16.6	86[4]	51.5
Iraq	370	49.2	14.8	26	52.6
Colombia	370	40.6	8.8	81	60ı9
Zambia	380	51.5	20.3	259[4]	44.5

Measures of health status by level of per capita Gross National Product (G.N.P.) in selected countries (continued)

Country	Per[1] capita G.N.P.	Crude[2] birth rate	Crude[2] death rate	Infant[3] mortality	Life[2] Expectancy
Guatemala	390	42.8	13.7	83	52.9
Malayasia	400	39.0	9.8	38	59.4
Dominican Republic	430	45.8	11.0	49	57.8
China, Republic of	430	26.7	10.2	18[7]	61.6
Iran	450	45.3	15.6	160[6]	51.0
Nicaragua	450	48.3	13.9	45[6]	52.9
Brazil	460	37.1	8.8	110[6]	61.4
Peru	480	41.0	11.9	67	55.7
Albania	480	33.4	6.5	87[4]†	68.6
Cuba	510	28.9	5.9	28	72.3
Costa Rica	590	33.4	5.9	56	68.2
Mexico	700	42.0	8.6	63	63.2
Jamaica	720	33.2	7.1	27	69.5
Portugal	730	18.4	10.1	50	68.0
Yugoslavia	730	18.2	9.2	44	67.5
Romania	740	19.3	10.3	40	67.2
Chile	760	25.9	8.1	71	64.3
Panama	820	36.2	7.1	34	66.5
Bulgaria	820	16.2	9.1	26	71.8
Hong Kong	900	19.4	5.5	17	70.0
Trinidad and Tobago	940	25.3	5.9	35[4]	69.5
Venezuela	1,060	36.1	7.0	52	64.7
Singapore	1,200	21.2	5.1	19	69.5
U.S.S.R.	1,400	17.8	7.9	23	70.4
Japan	2,130	19.2	6.6	12[4]	73.3
Israel	2,190	26.2	6.7	24	70.5
United States	5,160	16.2	9.4	19	71.3

Symbols: * for 1968
　　　　 † for 1965

Note:
Crude birth rates and death rates are births and deaths per 1,000 population per year. Infant mortality rate is number of deaths of children under one year of age per 1,000 live births per year. Life expectancy is expected length of life in years at birth.

Sources:
[1] World Bank. *World Bank Atlas*, 1973: 'Population, Per Capita Product and Growth Rates,' pp. 6–14. Washington: World Bank, 1973.
[2] United Nations projections, 1973. Unpublished data: averages for 1970–75.
[3] World Health Organisation. *The Fifth Report on the World Health Situation, 1969–72—Part II: Review by Country and Territory*, 'Population and Other Statistics,' by country, except where other sources are indicated. Unless otherwise noted, figures are for 1970–72. Geneva: WHO, 1974.
[4] United Nations. *Statistical Yearbook 1972*, Table 21, latest available year. New York: United Nations, 1973.
[5] World Health Organisation. *Malaria Control in Countries Where Time-limited Eradication is Impracticable at Present*. Report of a WHO Interregional Conference. WHO Technical Report Series No. 537, Annex 2, Table 2; figures are for 1971. Geneva: WHO, 1974.
[6] World Bank estimates, latest available year.
[7] United Nations. *Demographic Yearbook 1970*, Table 16; figure is for 1969. New York: United Nations, 1971.

Appendix VII

Official Aid

Bilateral aid from governmental sources in the industrialised countries accounts for 80 to 90 per cent of the total aid programmes. The particular objectives and the geographical areas involved in these aid operations vary from country to country. In the case of the U.K. the emphasis has been on the old colonies, and more recently, as a result of the interest of the present Minister, Mrs Judith Hart, on the needs of the very poor.

In addition to the bilateral aid programmes, other major aid programmes are coordinated and initiated by U.N. bodies and such programmes are known as multilateral programmes. There are other agencies, for example, the three regional development banks, Inter-American, African and Asian, and also the E.E.C. agencies. Finance is provided by the four U.N. agencies — the World Bank group, consisting of the International Bank for Reconstruction and Development (I.B.R.D.); the International Development Association (I.D.A.); and the International Finance Corporation (I.F.C.) together with the International Monetary Fund (I.M.F.), the three regional banks, and the two E.E.C. agencies, the European Development Fund, and the European Investment Bank.

Technical assistance is provided by the U.N. Development Programme (U.N.D.P.) which is financed by pledged contributions mainly from the U.S.A., Canada, Sweden, the U.K. and the Federal Republic of Germany. U.N.D.P. is concerned with the development of educational, training and research facilities. The U.N. Industrial Development Organisation (U.N.I.D.O.) was created to cover industrialisation. The Food and Agriculture Organisation (F.A.O.), the World Health Organisation (W.H.O.), the International Labour Organisation (I.L.O.) and the U.N. Educational, Scientific and Cultural Organisation (U.N.E.S.C.O.) initiate and coordinate programmes in their respective fields. The World Bank itself is also involved in technical programmes. Much of the funding for the specialist agencies comes from U.N.D.P.

Appendix VIII

Useful Addresses[1]

Regional Organisations

Appropriate Technology Centre
  Ministry of Finance, Planning and Economic Affairs, 17 B, Satellite Town, Rawalpindi, Pakistan
Appropriate Technology Development Association
  P.O. Box 311, Gandhi Bhawan, Lucknow, 226001 U.P., India
Appropriate Technology for Farmers Programme
  Institute of Agricultural Research, P.O. Box 2003, Addis Ababa, Ethiopia
Appropriate Technology Group
  9 Stanley Place, Minhana, Nugegode, Sri Lanka
Appropriate Technology Project
  Volunteers in Asia, Box 4543, Stanford, California 94305, U.S.A.
Arusha Appropriate Technology Pilot Project
  Small Industries Development Organisation, P.O. Box 764, Arusha, Tanzania
Christian Action for Development in Caribbean (C.A.D.E.C.)
  P.O. Box 616, Bridgetown, Barbados
Development Technology Centre
  Institute of Technology, P.O. 276, Jalan Ganesha 10, Bandung, Indonesia
Institute for Development Studies
  P.O. Box 30197, University of Nairobi, Kenya
International Centre of Tropical Agriculture (C.I.A.T.)
  Cali, Colombia
International Crops Research Institute for the Semi–Arid Tropics (I.C.R.I.S.A.T.)
  1–11–256, Begumpet, Hyderabad, 500006 A.P., India

International Institute of Tropical Agriculture (I.I.T.A.)
    Oyo Road, P.M.B. 5320, Ibadan, Nigeria
International Rice Research Institute (I.R.R.I.)
    P.O. Box 583, Manila, Philippines
Karen Village Technology Development and Demonstration
Unit
    U.N.I.C.E.F., P.O. Box 44145, Nairobi, Kenya
LikLik Buk Information Centre
    P.O. Box 1920, Lae, Papua, New Guinea
Tanzania Agricultural Machinery Testing Unit (T.A.M.T.U.)
    P.O. Box 1389, Arusha, Tanzania
Technology Consultancy Centre (T.C.C.)
    University of Science and Technology, Kumasi, Ghana
Technology Development and Advisory Unit (T.A.D.U.)
    University of Zambia, P.O. Box 2379, Lusaka, Zambia
Village Technology Unit
    Box 45, Jalan Halimun 4, Jakarta, Indonesia

International Organisations

Brace Research Institute
    MacDonald College of McGill University, Ste Anne De
    Bellevue, Quebec, Canada HOA 1CO
C.I.I.R. Overseas Volunteers
    1 Cambridge Terrace, London NW1 4JL
Intermediate Technology Development Group Ltd (I.T.D.G.)
    9 King Street, Covent Garden, London WC2E 8HN
International Voluntary Service
    53 Regent Road, Leicester LE1 6YL
Stichting Tool (T.O.O.L.)
    Mawitskade 61a, Amsterdam. Netherlands
Tropical Products Institute (T.P.I.)
    56–62 Grays Inn Road, London WC1X 8LU
United Nations Association — International Service
    3 Whitehall Court, London SW1A 2EL
Voluntary Service Overseas
    14 Bishops Bridge Road, London W2 6AA
Volunteers in Technical Assistance (V.I.T.A.)
    3706, Rhode Island Avenue, Mt Rainer, Maryland 20822,
    U.S.A.

References

Chapter 1

1.   Schumacher, E. F., *Small is Beautiful* (Sphere Books, London, 1974) (Harper & Row, New York, 1973 [rev. eds. 1975 & 1976]).
2.   *Appropriate Technology: Problems and Promises*, ed. N. Jéquier (O.E.C.D., Paris, 1976).
3.   Fraenkel, P., *Food From Windmills* (Intermediate Technology Publications, London, 1976).
4.   Bateman, G., *The Introduction of Rainwater Catchment Tanks and Micro-Irrigation to Botswana* (Intermediate Technology Publications, London, 1969).
5.   Stettner, L., 'An Intermediate Approach to Low-Cost Housing In Ghana', *Appropriate Technology*, vol. I, no. 1.
6.   Freire, P., *Cultural Action for Freedom* (Penguin Education, 1972) (*Harvard Educational Review* [Monograph Series: No. 1], Cambridge, Mass.).
7.   Marsden, K., *Progressive Technologies for Developing Countries* (Small Industries Unit, I.L.O., 1967).

Chapter 2

1.   Myrdal, G., *Economic Theory and Underdeveloped Regions* (Gerald Duckworth, London, 1957).
2.   Bauer, P. T., *Dissent on Development* (Weidenfeld and Nicolson, London, 1971) (Harvard University Press, Cambridge, Mass., 1972 [pb. 1976]).
3.   Rostow, W. W., *The Stages of Economic Growth, A Non-Communist Manifesto* (Cambridge University Press, 1960).
4.   Schiavo-Campo, S., and Singer, H. W., *Perspectives of Economic Development* (Houghton Mifflin, New York, 1970).
5.   Lewis, W. A., *Theory of Economic Growth* (Homewood, Illinois, 1955).
6.   Todaro, M. P., *Economics For a Developing World* (Longman, London, 1977).
7.   Donaldson, P., *Worlds Apart: The Economic Gulf Between Nations* (Penguin, London, 1973).
8.   Kaldor, N., *Industrialisation in Developing Countries*, edited by R. Robinson (Cambridge University Overseas Studies Committee, Cambridge, 1965).
9.   Schumacher, E. F., *Small is Beautiful* (Sphere Books, London, 1974) (Harper & Row, New York, 1973 [rev. eds. 1975 & 1976]).
10. Meadows, D. H., Meadows, D. L., Randers, J., and Behrens, W. W., *The Limits to Growth* (Earth Island, London, 1972) (New American Library, New York).

11. Mesarovic, M., and Pestel, E., *Mankind at The Turning Point* (Hutchinson, London, 1975) (New American Library, New York, 1976).
12. Kahn, H., *The Next 200 Years* (W. Morrow & Co., New York, 1976).
13. Chapman, P., *Fuels Progress* (Penguin, London, 1975).

Chapter 3

1. Makhijani, A., and Poole, A., *Energy and Agriculture in the Third World* (Ballinger Publishing Company, Cambridge, Mass., 1975).
2. Vaughan, B., *Doctor in Papua* (Hale & Co., London, 1974).
3. Thring, M. W., and Laithwaite, E. R., *How to Invent* (Macmillan Press, London, 1977).
4. Papanek, V., *Design for the Real World* (Thames and Hudson, London, 1971) (Bantam, New York, 1976).
5. Congdon, R. J., *Student Projects Briefing Pamphlet* (Intermediate Technology Publications, London).
6. McRobie, G., 'The Mobilisation of Knowledge on Low Cost Technology: Outline of a Strategy', *Appropriate Technology: Problems and Promises*, ed. N. Jéquier (O.E.C.D., Paris, 1976).
7. I.T.D.G. Publications Ltd, 9 King Street, London WC2E 8HN.
8. *Appropriate Technology*, vol. 1 (Appropriate Technology Association, Post Box 311, Gandhi Bhawan, Lucknow, India, 1977).
9. *Village Technology Handbook* (V.I.T.A., Mount Rainier, Maryland, U.S.A., 1975).
10. *Handbook of Appropriate Technology* (Brace Research Institute, Quebec, 1975).
11. *Liklik Buk. A Rural Development Handbook: Catalogue for Papua, New Guinea, 1977* (Wantok Publications, P.O. Box 1982, Boroko, Papua, New Guinea).
12. Fuglesang, A., *Applied Communication in Developing Countries* (Dag Hammarskjold Foundation, 1973).
13. Bowers, *et al.*, *Action Research and the Production of Communication Media* (printed by the Typography Unit, Reading University, 1973).

Chapter 4

1. Gordon, J. E., *The New Science of Strong Materials* (Penguin, London, 1968; New York, 1975).
2. Miller, G. T., *Energetics, Kinetics and Life* (Wadsworth, California, 1971).
3. Aylward, F., and Jul, M., *Protein and Nutrition Policy in Low Income Countries* (Charles Knight, London, 1975) (Halsted Press, New York, 1975).
4. *Liklik Buk. A Rural Development Handbook: Catalogue for Papua, New Guinea, 1977* (Wantok Publications, P.O. Box 1982, Boroko, Papua, New Guinea).
5. Pirie, N. W., *Food Resources, Conventional and Novel* (Penguin, London, 1976).

6. George, S., *How the Other Half Dies* (Penguin, London, 1976) (Universe Books, New York, 1977).
7. Finney, C., *Farm Power in West Pakistan* (University of Reading, Development Study No. 11, 1972).
8. *Green Revolution?*, ed. B. H. Farmer (Macmillan, London, 1977) (Westview Press, Boulder, Colorado, 1977).
9. Carr, M., 'Simple Technologies for Villages in Africa', *Proceedings of Conference on the Effective Use of Appropriate Technologies* (Lilly Endowment Inc., April 1977).
10. Wijewardene, R., private communication.
11. *Tools for Agriculture: A Buyer's Guide to Low-Cost Agricultural Implements*, compiled by J. Boyd (Intermediate Technology Publications, London, 1976) (International Scholarly Book Services, Forest Grove, Oregon, 1977).
12. Macpherson, G. A., *First Steps in Village Mechanisation* (Tanzania Publishing House, Dar Es Salaam, 1975).
13. Minto, S. D., and Westley, S. B., *Low-Cost Rural Equipment Suitable for Manufacture in East Africa* (Institute of Development Studies, P.O. Box 30197, University of Nairobi, Kenya, 1975).
14. *Oil Drum Forges* (Intermediate Technology Publications, London, undated).
15. *Village Technology Handbook* (V.I.T.A., Mount Rainier, Maryland, U.S.A., 1975).
16. *Multi-Row Weeder* (Intermediate Technology Publications, London, undated).
17. *The Weeder Mulcher* (Intermediate Technology Publications, London, undated).
18. Boyd, J., 'Tools for Agriculture', *Lectures on Socially Appropriate Technology* (Technische Hogeschool, Eindhoven, Netherlands, 1975).
19. Oke, O. L., Olatunbosun, D. A., and Adadevoh, B. K., 'Leaf Protein: A New Protein Source for the Management of Protein Calorie Malnutrition in Nigeria', *Nigerian Medical Journal*, vol. 2, no. 4 (October 1972).

Chapter 5

1. Stern, P. H., 'Low Cost Development of Water Services', *Proceedings of Seminar on Appropriate Technology in Economic Development* (University of Edinburgh, September 1973).
2. Tinker, Jon, 'Nor Any Drop to Drink', *New Scientist* (17 March 1977).
3. *Water, Waste and Health in Hot Climates*, ed. R. Feachem *et al.* (Wiley, London and New York, 1977).
4. Mann, H. T., and Williamson, D., *Water Treatment and Sanitation* (Intermediate Technology Publications, revised edition, London, 1976) (International Scholarly Book Services, Forest Grove, Oregon, 1977).
5. *Water for the Thousand Millions*, ed. A. Pacey (Pergamon, Oxford and New York, 1977).
6. Watt, S. B., and Wood, W. E., *Hand Dug Wells and their Construction* (Intermediate Technology Publications, London, 1976) (International Scholarly Book Services, Forest Grove, Oregon, 1977).
7. *Village Technology Handbook* (V.I.T.A., Mt Rainier, Maryland, U.S.A., January 1977).

8.   Molenaar, A., *Water Lifting Devices for Irrigation* (F.A.O., Rome, 1956).
9.   Minto, S. D., and Westley, S. B., *Low-Cost Rural Equipment Suitable for Manufacture in East Africa* (Institute for Development Studies, P.O. Box 30197, University of Nairobi, Kenya, 1975).
10. Dickinson, H., and Winnington, T. L., 'Rural Technology in China', *Appropriate Technology*, vol. 1, no. 1 (1973).
11. Zapp, J., *Intermediate Technology* (University of Los Andes, Bogota, Colombia, 1974).
12. Fraenkel, P., *Food from Windmills* (Intermediate Technology Publications, London, 1976) (International Scholarly Book Services, Forest Grove, Oregon, 1977).
13. Watt, S., *A Manual on the Automatic Hydraulic Ram Pump* (Intermediate Technology Publications, London, 1975).
14. Humphrey, H. A., 'An Internal Combustion Pump, and other applications of a New Principle', *Proceedings of Institute of Mechanical Engineers* (December 1909).
15. Dunn, P. D., 'The Humphrey Pump for use in Developing Countries', *Proceedings of Seminar on Appropriate Technology in Economic Development* (University of Edinburgh, September 1973).
16. Belcher, A. E., *The Hydrostatic Pump* (Micro-D Ltd, 344–350 Euston Road, London NW1 3BJ, 1972).
17. MacCracken, C. D., 'The Solar Powered Thermopump' *Transcript of Conference on Solar Energy* (Tucson, Arizona, 1955).
18. Girardier, J. P., Alexandroff, M. G., and Alexandroff, J. M., 'Les moteurs solaires et l'habitat pour les zones arides: réalisations actuelles et perspectives' *Proceedings of International Congress, The Sun in the Service of Mankind* (UNESCO, Paris, July 1973).
19. *Appropriate Technology*, vol. 1 (Appropriate Technology Association, Post Box 311, Gandhi Bhawan, Lucknow, India, 1977).

Chapter 6

1.   Obayazid, O. M., *Prospects of Fuel and Energy in the Sudan* (Council for Scientific and Technological Research, C.S.T.R./12/D/3/8, Khartoum, 1975).
2.   *Water for the Thousand Millions,* ed. Arnold Pacey (Pergamon, Oxford and New York, 1977).
3.   Bott, A. N. W., 'Power Plus Proteins from the Sea', *Journal of Royal Society of Arts* (July 1975).
4.   Levy, J. P., 'Distillation, Pyrolysis and Alcohol Production from Wood' *Proceedings of International Conference on Solar Energy in Agriculture* (Reading, 1976, I.S.E.S., London).
5.   Hamid, Y. G., 'An Experimental Solar Still Design for the Sudan', *Appropriate Technology,* vol. 3, no. 3 (1976).
6.   Hoda, M. M., 'Solar Cooker' *Proceedings of UNESCO International Conference on Solar Building Technology* (London, July 1977).
7.   Wilson, S. S., 'Pedal Power', *Lectures on Socially Appropriate Technology* (Technische Hogeschool, Eindhoven, Netherlands, 1975).
8.   Rao, D. P., and Rao, K. S., 'Solar Pump for Lift Irrigation', *Solar Energy*, vol. 18 (Pergamon, Oxford, 1976).
9.   Girardier, J. P., Alexandroff, M. G., and Alexandroff, J. M., 'Les

moteurs solaires et l'habitat pour les zones arides: réalisations actuelles et perspectives' *Proceedings of International Congress, The Sun in the Service of Mankind* (UNESCO, Paris, July 1973).

10. West, C., 'The Fluidyne Heat Engine, Harwell, England', U.K.A.E.R.E. *Research Report* R. 6775 (1974).
11. Parkes, M. E., 'The Use of Windmills in Tanzania' (Paper No. 33, June 1974, Bureau of Resource Assessment and Land Use Planning University of Dar es Salaam, Tanzania).

Chapter 7

1. Harrison, P., 'Basic Health Delivery in the Third World', *New Scientist* (17 February 1977).
2. Smith, A. J., 'Medicine in China: Barefoot Doctors and the Medical Pyramid', *Appropriate Technology*, vol. 1, no. 3 (Autumn 1974).
3. Dickinson, H., 'Rural China, 1972', Report for World Council of Churches (1972).
4. *The Training of Auxiliaries in Health Care*, compiled by Katherine Elliott (Intermediate Technology Publications, London, 1975) (International Scholarly Book Services, Forest Grove, Oregon, 1977).
5. El Agil, A. A. R., and Erwa, H. H., 'Decontamination of Foodstuffs by Solar Energy: Bacterial Counts in Food Samples following exposure to sunlight in airtight containers', I.R.C.S. 2, 1270 (1974).
6. Davies, M., 'I.T. Suburban Style in Africa', *Appropriate Technology*, vol. 3, no. 1 (1976).
7. *Appropriate Technology*, vol. 1 (Appropriate Technology Association, Post Box 311, Gandhi Bhawan, Lucknow, India, 1977).
8. *Simple Designs for Hospital Equipment* (12 leaflets produced by Intermediate Technology Publications, London).
9. Parry, J. P. M., 'Intermediate Technology Building', *Lectures on Socially Appropriate Technology* (Technische Hogeschool, Eindhoven, Netherlands, 1975).
10. Griffiths, J. F., *Applied Climatology: An Introduction* (Oxford University Press, London and New York, 1966) (2nd ed., 1976).
11. Spence, R. J. S., 'Building Materials for Rural Areas: The Need for Research', *Proceedings of Seminar on Appropriate Technology in Economic Development* (University of Edinburgh, 1973).
12. Spence, R. J. S., 'Laurie Baker and the Technology of Low Cost Building', *Appropriate Technology*, vol. 1, no. 3 (Autumn 1974).
13. Thyagarajau, G., private communication.
14. Bahadori, M. N., 'Solar Energy at Pahlavi University', *Sun World*, I.S.E.S. no. 3 (February 1977).
15. *Carts* (I.T.D.G. Publications, London, undated).
16. *How to Make a Metal Bending Machine* (Intermediate Technology Publications, London, 1973).
17. Dickinson, H., and Winnington, T. L., 'Ferro-cement for Boat Building', *Proceedings of Seminar on Appropriate Technology in Economic Development* (University of Edinburgh, 1973).

Chapter 8

1. CoSIRA, 35 Camp Road, Wimbledon, London SW19.

2.   'The Development and Operation of Graham Centre' *Proceedings of Conference on the Effective Use of Appropriate Technologies* (Lilley Endowment Inc., April 1977).
3.   Ntim, B. A., and Powell, J. W., 'Appropriate Technology in Ghana: The Experience of Kumasi University's Technology Consultancy Centre', *Appropriate Technology, Problems and Promises*, ed. Nicolas Jéquier (O.E.C.D. Publications, 1976).
4.   Hoda, M., 'India's Experience and the Gandhian Tradition', *Appropriate Technology, Problems and Promises* (O.E.C.D. Publications, 1976).
5.   Harper, M., *Partnership for Productivity* (I.T.D.G. Publications, May 1977).

Chapter 9

1.   Tett, C. R., 'Education Systems: Appropriate Education and Technology for Development', *Lectures on Socially Appropriate Technology* (Technische Hogeschool, Eindhoven, Netherlands, 1974).
2.   'Educational Development, World and Regional Statistical Trends and Projections until 1985' (a UNESCO background paper prepared for the World Population Conference, Bucharest, 1974).
3.   Coombs, P. H., Prosser, R. C., and Ahmed, M., *New Paths to Learning for Rural Children and Youth* (New York International Council for Educational Learning, 1973).
4.   Dunn, P. D., 'Engineers for A.D. 2000', *Education* (23 August 1968).

Appendix II

1.   Samuelson, P. A., *Economics*, 10th edn (McGraw-Hill, New York, 1976).

Appendix III

1.   Congdon, R. J., *Student Projects Briefing Pamphlet* (Intermediate Technology Publications, London).

Appendix IV

1.   *Water, Waste and Health in Hot Climates*, ed. R. Feachem, M. McGarry and D. Mara (Wiley, London and New York, 1977). Much of the material for this appendix has been taken from this book, particularly from chapters 1, 2 and 5.
2.   'Working Capacity of Bilharzia Patients', *Appropriate Technology*, vol. 1, no. 2 (Summer 1974).
3.   *Village Technology Handbook* (V.I.T.A., Mt Rainier, Maryland, U.S.A., 1975).

Appendix VI

1.   *The Assault on World Poverty* (Johns Hopkins Press for the World Bank, Baltimore, 1975).

Appendix VIII

1. *Appropriate Technology in the Commonwealth: A Directory of Institutions* (Commonwealth Secretariat, Marlborough House, Pall Mall, London SW1Y 5HX). This book provides a comprehensive list of organisations in the appropriate technology field. It gives addresses and a brief account of the range of activities of the organisations.

Further Reading

Chapter 1

Schumacher, E. F., *Small is Beautiful* (Sphere Books, London, 1974) (Harper and Row, New York, 1973 [rev. eds. 1975 and 1976]).
McRobie, G., *Small is Possible* (Harper and Row, New York, in the press).

Chapter 2

Historical background to economic development

> Cipolla, C. M., *Before The Industrial Revolution: European Society and Economy 1000–1700* (Methuen, London, 1976) (W. W. Norton & Co., New York). Provides a very good introduction to European development before the Industrial Revolution.
> Rostow, W. W., *How It All Began: Origins of the Modern Economy* (Methuen, London, 1976) (McGraw-Hill, New York, 1975). The author expands his ideas on the stages of growth theory.

General economics

> Samuelson, P. A., *Economics* (10th edn, McGraw-Hill, New York, 1976). An excellent introductory text book, and the fact that it is now in its 10th edition indicates the regard in which this book is held.
> Galbraith, J. K., *Economics, Peace and Laughter* (Penguin, London, 1971) (New American Library, New York, 1972). A book which is both witty and wise.

Development economics

> Donaldson, P., *Worlds Apart: The Economic Gulf Between Nations* (Penguin, London, 1973), and Elkan, W., *An Introduction to Development Economics* (Penguin, London, 1973). Provide a simple introduction to the subject.
> McQueen, M., *The Economics of Development* (Weidenfeld and Nicolson, London, 1973). A particularly good introductory book.
> Todaro, M. P., *Economics for a Developing World* (Longman, London, 1977). An up to date and comprehensive book written primarily for developing countries.
> Higgins, B., *Economic Development. Principles, Problems and Policies* (Constable, London, 1959) (W. W. Norton & Co., New York [rev. ed. 1968]. This book provides a comprehensive treatment of the more important economic development theories. The non-specialist may find it somewhat difficult.
> Bauer, P. T., *Dissent on Development* (Weidenfeld and Nicolson, London, 1971) (Harvard University Press, Cambridge, Mass., 1972

[pb. 1976]). An outstanding work. Clear, comprehensive in its coverage and forthright and controversial in style. This lively book makes compulsive reading.

Futurology

A *Blueprint for Survival* (Penguin, London, 1972) (New American Library, New York). A pioneering work in the field. Draws attention to the problems of pollution and resource depletion and refers to the dangers resulting from exponential growth.

Taylor, G. R., *The Doomsday Book* (Thames and Hudson, London, 1970). An early and influential book which expresses concern on the problems of pollution.

Meadows, D., *et al.*, *The Limits to Growth* (Earth Island, London, 1972) (New American Library, New York). Describes the results of a multiloop feedback model of the world. The results predict catastrophic effects for the future.

Mesarovic, M., and Pestel, E., *Mankind at the Turning Point* (Hutchinson, London, 1975) (New American Library, New York, 1976). Extends the Meadows model and breaks down the world into ten groups.

Allaby, M., *Who Will Eat?* (Tom Stacey, London, 1972). A readable account of future trends with particular reference to world food problems.

Kahn, H., *The Next 200 Years* (W. Morrow and Co., New York, 1976). Sets out the 'Technological Fix' alternatives. Should be read in conjunction with references 10 and 11 to get both sides of the picture.

Maddox, J., *The Doomsday Syndrome* (McGraw-Hill, London and New York, 1972) and Cole, H., *et al.*, *Thinking About the Future* (Chatto and Windus, London, 1973). Both books provide a critical appraisal of the Meadows methods and conclusions, and should be read in order to obtain a better overall assessment of the problems.

Chapter 3

The section by Jéquier in *Appropriate Technology: Problems and Promises*, ed. N. Jéquier (O.E.C.D., Paris, 1976) provides an interesting discussion on A.T. and its origins.

The processes of invention and design are described in *How to Invent* by M. W. Thring and E. R. Laithwaite (Macmillan Press, London, 1977), a book whose two authors are both distinguished engineers and inventors.

A lively personal view of innovation, with many examples, is given in V. Papanek, *Design for the Real World* (Thames and Hudson, London, 1971).

The subject of communication is introduced in A. Fuglesang, *Applied Communication in Developing Countries* (Dag Hammerskjold Foundation, 1973). This is an outstanding book and should be more widely known.

Useful design data and suppliers' addresses are given in the handbooks, *Appropriate Technology*, vol. 1 (Appropriate Technology Association, Post Box 311, Gandhi Bhawan, Lucknow, India, 1977); *Village Technology Handbook* (V.I.T.A., Mount Rainier, Maryland, U.S.A., 1975); *Handbook of Appropriate Technology* (Brace Research Institute, Quebec, 1975); *Liklik Buk, A Rural Development Handbook: Catalogue for Papua, New Guinea, 1977* (Wantok Publications, P.O. Box 1982, Boroko, Papua, New Guinea). I.T.D.G. Publications publish a number of useful handbooks. The I.T.D.G.

journal *Appropriate Technology* reports on current developments and gives design information.

Economically Appropriate Technologies for Developing Countries: An Annotated Bibliography, compiled by Marilyn Carr (Intermediate Technology Publications, London, 1976) (International Scholarly Book Services, Forest Grove, Oregon, 1977). Provides an extensive list of publications in the A.T. field together with a short summary of the contents.

Chapter 4

Borgstrom, G., *World Food Resources* (International Text Book Co., 1973) (T. Y. Crowell, New York, 1973). This book provides a good general introduction to world food production and consumption. It also discusses nutrition, food technology, and the energy requirements of agriculture.

Pyke, M., *Man and Food* (World University Library, 1970) (McGraw–Hill, New York, 1970). Provides a lively introduction to food nutritional requirements and food technology.

Pirie, N. W., *Food Resources, Conventional and Novel* (Penguin, London, 1976). A clearly written and authoritative account of new approaches to the problem of satisfying future food demands.

Aylward, F., and Jul, M., *Protein and Nutrition Policy in Low Income Countries* (Charles Knight, London, 1975) (Halstead Press, New York, 1975). A comprehensive survey of the protein and nutritional problems and policies in low income countries. The book is based on a report of the Protein Advisory Group set up by the U.N.

Culpin, C., *Farm Machinery* (9th edn, Crosby Lockwood Staples, London, 1976) (Beekman Pubs., New York, 1977). A standard textbook on agricultural engineering. The Handbooks given in references 8, 9, 10 and 11 of chapter 3 provide information on designs of A.T. agricultural equipment.

Tools for Agriculture: A Buyer's Guide to Low-Cost Agricultural Implements, compiled by J. Boyd (Intermediate Technology Group Publications, London, 1976) (International Scholarly Book Services, Forest Grove, Oregon, 1977). This book lists a considerable number of items of low-cost agricultural equipment and gives details of their characteristics and the names and addresses of manufacturers. It replaces the earlier *Tools For Progress* published by I.T.D.G. (Additional publications are available from I.T.D.G. describing the design and manufacture of low-cost agricultural equipment.)

Chapter 5

Water for the Thousand Millions, ed. A. Pacey (Pergamon, Oxford and New York, 1977), provides a readable introduction to the subject of low-cost water supplies.

Allsebrook, J. C. P., 'Where shall we dig the Well?', *Appropriate Technology*, vol. 4, no. 1 (May 1977). This article explains the elementary geological considerations underlying site selection and provides a valuable introduction to the subject.

Village Technology Handbook (V.I.T.A., Mount Rainier, Maryland, U.S.A., January 1977) gives a detailed account of boring and well digging.

The construction of hand–dug wells is described in *Hand Dug Wells and their Construction*, S. B. Watt and W. E. Wood (Intermediate Technology Publications, London, 1976) (International Scholarly Book Services, Forest Grove, Oregon, 1977).

Water treatment methods are set out in *Water Treatment and Sanitation* by H. T. Mann and D. Williamson (Intermediate Technology Publications, revised edition, 1976).

An excellent and thorough treatment of all aspects of water, waste and health is provided by *Water, Waste and Health in Hot Climates*, ed. R. Feachem *et al.* (Wiley, London and New York, 1977).

Molenaar gives a comprehensive survey of traditional and small–scale modern pumping equipment. Molenaar, A. *Water Lifting Devices for Irrigation* (F.A.O., Rome, 1956).

Hudson, N. W., *Field Engineering for Agricultural Development* (Oxford University Press, Oxford and New York, 1975). A good introductory textbook on water management and irrigation.

Water for Arid Lands: Promising Technologies and Research Opportunities (National Academy of Sciences, Washington D.C., 1974). A number of new technologies on water conservation and use are described in this publication.

Chapter 6

Energy for Rural Development. Renewable Resources and Alternative Technologies for Developing Countries (National Academy of Sciences, Washington, D.C., 1976). Provides a good state of the art survey.

Makhijani, A., and Poole, A., *Energy and Agriculture in the Third World* (Ballinger, U.S.A., 1975). A valuable study.

The Energy Primer (Compendium, 234 Camden High Street, London NW1). A useful source of ideas and references.

Guide to Small Power Sources (I.T.D.G. Publications, London, 1978). Lists a large number of suppliers of low–cost power converters including windmills, solar devices, and small internal combustion engines.

Wind power

Golding, E. W., *Generation of Electricity by Wind Power* (E. and F. N. Spon, London, 1976) (Halsted Press, New York, 1976). This classic work has now been reprinted with an additional chapter on recent developments in wind power research. The book is particularly good on measurement and site selection.

Park, J., *Simplified Wind Power Systems for Experimenters* (Helion, Box 4301, Sylmar, California 91342, 1976). A very useful, practical introduction to the design of small windmills.

Solar energy

Daniels, F., *Direct Use of the Sun's Energy* (Ballantine Books, New York, 1975). A classic by an early American solar energy authority. This book, first published in 1964, is well worth reading.

Brinkworth, B. J., *Solar Energy for Man* (Compton Press, 1972) (Halsted Press, New York, 1973). A clear treatment of solar energy and its conversion.

Methane

Proceedings of an I.T.D.G. Seminar (Intermediate Technology Publications, London, 1976) (International Scholarly Book Services, Forest Grove, Oregon, 1977). A useful introduction.

Fry, J. L., *Practical Building of Methane Power Plants for Rural Energy Independence*
ed. D. A. Knox (Andover, Hampshire, 1975) (L. John Fry, Santa Barbara,
California, 1974). An excellent practical textbook by an author with extensive
personal experience of the construction of methane generators.
Meynell, P.-J. *Methane: Planning a Digester* (Schocken Books, New York,
1978). A detailed, easy-to-follow handbook on how to build a digester to
produce methane.

Engines
Ayres, R., and McKenna, R., *Alternatives to the Internal Combustion Engine*
(Johns Hopkins Univ. Press, U.S.A.). Though written primarily from the
point of view of alternatives to the automative petrol engine, this book provides
an excellent survey of the characteristics of the internal combustion engine and
the possible alternatives to it. Descriptions of the Stirling and steam engines are
included.

Pedal power
Pedal Power: In Work, Leisure and Transportation, ed. J. C. McCullogh (Rodale
Press, Emmaus, Pa., 1977).

Chapter 7

Medical care and services
The Training of Auxiliaries in Health Care, compiled by Katharine Elliott
(Intermediate Technology Publications, London, 1975) (International Scholarly Book Services, Forest Grove, Oregon, 1977) is a useful annotated bibliography.

The Macmillan *Tropical Community Health Manuals* comprise a series of
short, practical books addressed to practising nurses, auxiliaries, medical
officers and assistants and others engaged in rural health centres and sub-
centres. Titles include

> Ebrahim, G. J., *Handbook of Tropical Paediatrics*
> Ebrahim, G. J., *Care of the Newborn in Developing Countries*
> Ebrahim, G. J., *Child Care in the Tropics*
> Ebrahim, G. J., *Breast feeding — the Biological Option*
> Ebrahim, G. J., *Practical Mother and Child Health in Developing Countries*

Designs for hospital equipment are given in the I.T.D.G. leaflets in reference 8.

Building
Turner, J. F. C., *Housing by People* (Marion Boyars, London, 1976) (Pantheon,
New York, 1977). An unusual book of the 'Illich' type by an architect with
extensive experience in low-cost housing in South America and elsewhere.
Dancy, H. K., *A Manual on Building Construction* (I.T.D.G. Publications,
London, 1975) (International Scholarly Book Services, Forest Grove,
Oregon, 1977). The book was originally published in 1948 by Rev. Dancy,
of the Sudan Interior Mission, and entitled 'Mission Building'. It is a mine of
information on simple practical methods of low-cost building.

Also useful is
Miles, D., *A Manual of Building Maintenance* (vol. 1 *Management,* vol. 2 *Methods*)
(I.T.D.G. Publications, London, 1976) (International Scholarly Book Ser-
vices, Forest Grove, Oregon, 1977).

Chapter 9

Illich, I., *Deschooling Society* (Penguin, London, 1973) (Harper & Row, New York, 1972). The author questions whether schooling is the same as education. The book raises a number of fundamental questions on the value of traditional education.

Index

Acupuncture 134
Afghan joggle pump 102
Africa 65, 67, 69, 91
Agricultural Equipment, Local manufacture of 72; *see also* cultivators, harrows, levellers, ploughs, seeders, weeders
Agricultural sector 28, 29
Agriculture, Modern 67
Agricultural processes 74
Aid, Official 28, 176, 198
Aid, Bilateral 198
Aid, Multilateral 198
Air lift pump 100
Airports 42
Air tower 141–2
Alcan 108
Algae, blue green 56
Allsebrook, C. 91
Alternative societies 40, 46
Animal 58; as internal combustion engine 123
Anvil 72–3
'Applied Communication in Developing Countries' 48
Appropriate Technology v, vii, *et seq*, 1 *et seq*, 3, 29, 39, 40 *et seq*, 44, 53, 174; aims of 5, 41; centres 5, 48, 155; claims of 41; courses in 170; equipment, examples of 6 *et seq*; failures of 43; objections to 41; research and development 171
Appropriate Technology Development Association, A.T.D.A. 155, 175
Appropriate Technology Group, A.T.G.S.L. Sri Lanka 155
'Appropriate Technology' Journal 48, 174
Arbab, F. 163
Asia 65–6
Aswan Project 42, 90
Automation 157

Bacon, Francis xi
Backwash argument 26
Baffoe, S. K. 152
Bagasse 110
Baker, L. 140
Banks, regional development 147
Bare foot doctors 13, 44, 133, 162
Barometer 99
Basic human requirements 5
Batteries 129
Bauer, P. T. 26, 28

Bellows 72
Beri-beri 63
Biafra, innovation in 46
Bicycle, use of parts 128; *see also* dynapod, power 122
'Big push' 6
Bilharzia, *see* schistosomiasis
Bio-gas (methane) 43, 103, 117, 123
Bio-mass (vegetation, organic matter) 116
Births 133
Birth control 33, 134
Birth rate 30, 134
Boats, reinforced concrete 144–5
Bore holes (tube wells) 93; casing 94; number out of action due to pump breakdown 115; vandalism of 91
Bottlenecks, labour 70
Botswana 9, 166
Bott, A. N. W. 116
Bowers, J. 50
Brace Research Institute 5, 48, 175
Bradley, D. J. 87, 185 *et seq*
Brayton Cycle 120
Brazil 118
Bricks fired clay 139; stabilised soil 138–9
Bucket, infection from 87, 93
Building 136 *et seq*; heating and cooling 141; reasons for change 138; traditional 136–7

California 46
Calorie 19, 61
Capital formation 28
Capitalist system 26, 41
Carbohydrates 56, 61
Carr, M. 69
Cassava (manioc) 66, 118; grinder 124
Catchment tanks 9, 108
Cauca, River 76
Cement Portland 141, 155; Pozzolana 141
Centrally Planned Economy 23
Cereal consumption 59
Chapman, P. 38
Charcoal 119
Charka 4
Chesterton, G. K. 177
China 4, 13, 44, 133, 136, 147
Chlorination 89
Cholera 87, 134, 185 *et seq*
CINVA Ram 139
Clothing 5, 58
Club of Rome 32 *et seq*

Coal 116
Colombia 18, 20–3, 42, 44, 76, 102, 128, 133, 134, 161, 163
Colonial exploitation 22
Communication 46
Complexity of operation 115
compression ignition engine, *see* diesel engine
Congdon, R. J. 46
Conversion factors 179
Cooking, traditional 113, 114, 117, 119; solar 122, 124, 166
Cost of energy 115
Council for Small Industries in Rural Areas, Co.S.I.R.A. 149
Crop yield 69; losses 69; processing 114
Cultivation 77
Cultivators 78

De Bono, lateral thinking 45
De Groot, Rijn 157
Desalination 114
Development 22; aims 5; centre 150; terms of reference 150; economists 22; gap 23
Developing countries 1
Developing market economy 23
Dickens, Charles 32
Dickenson, H. 170
Diesel engine 103, 123
Digester (fermentor) 117–8
Directory of Appropriate Technology 48
Disguised unemployment 27
Doctors, bare foot 13, 44, 133, 162; distribution of 133; per capital 21, 133
Doors 140–1
Doubling time 30
Do it yourself industry 45, 166
Dualism 3, 19, 25
Dual society 3
Dung 117
Dynapod 122
Dysentery 5.1, 134, 185 *et seq*

East Africa 73, 79
Ecological damage 32
Econometrics 27
Economists, Development 22, 25
Economic growth 22
Economic indicators 19
Education 158 *et seq*
Education, minimum learning package 160
Education, Informal 13, 160; primary 159, 166; secondary 167; traditional 160; university 167 *et seq*; western 160; statistics 192 *et seq*
Educational equipment 167, 170
E.E.C. 198
Egg tray manufacture 14–6
Egypt 90, 105
Electric motor 103, 120
Electricity generation 113
Electrification, Rural 42
Electron gas 120
Elephantiasis 88, 185 *et seq*
Emerging Countries 1
Energy 111 *et seq*, 190 *et seq*; accounting 38;
consumption 111; consumption per capita 23, 111 *et seq*; consumption in the Sudan 114; consumption in the U.S.A. 114; cooking 114, 119; equivalent of food 111; units of measurement 179; resources 115; solar desalination 120–1
Energy converters 119; *see also* diesel engine, fluidyne, gas cycle, gas turbine, heat engine, hydraulic engine, internal combustion engine, muscle power, petrol engine, photocell, reciprocating engine, solar powered engine, steam engine, Stirling engine, thermocouple, vapour cycle engine, water wheel, water turbine, windmill
Energy sources 115; *see also* biomass, charcoal, coal, fossil fuel, firewood, geothermal, hydropower, natural gas, nuclear, oil, solar, tidal, wave
Energy sources, selection criteria 114
Energy, Use in rural areas 114
Energy, Use in the Sudan 114
Energy, Use in the U.S.A. 114
Engineering drawing 50–1
Engineers, Training of 168 *et seq*
Entrepreneurs 146 *et seq*
Environmental effects 32
Erosion 76
Ersatz 46
Ethiopia 7, 8, 18, 20–3
Ethnic differences 22
European Development Fund 198
European Investment Bank 198

Family limitation 134
Far East 64, 76
Farmers, Large and the Green Revolution 68
Farmers, Small and the Green Revolution 68
Farmers, Small, problems of 70
Fats 61
Feachem, N. 185
Fermentation 67
Fermentation to alcohol 117
Fertiliser needs 68, 75; plant requirements 57, 58; per capita use 22
Fibre glass 145
Filter 89
Firewood 113, 119
First law of thermodynamics 190
Fish farming 66
Flap valve pump 162, 163
Fluidyne 127
Food basic crops 66; chain 59, 60; consumption per capita 19; energy equivalent per capita 111; human nutritional requirements 61 *et seq*; legumes 66; pyramid 60; web 60; world food supplies 65
Forge simple design 72–3; I.T.D.G. design 73–4
Fossil fuel 115
Freire, P. 13
Friction head 95
Fuglesang, A. 48, 50
Fund raising 176
Fungicides 58, 68, 75

Fusion nuclear 116
Futuro, Colombia 52, 161
Futurologists 32 *et seq*

Gandhi, M. 4
Gandhian Institute of Studies 48
Gas cycle 120
Gas turbine 120
Gastro enteritis 88, 185 *et seq*
Geothermal power 115
Ghana 5, 11, 48, 139, 152; *see also* Kumasi
Goats, damage by 10
Green Revolution 44, 68
Gross Domestic Product, G.D.P. 181
Gross National Product, G.N.P. 2, 19 *et seq*
Growth 1, 22
Guinea worm 87, 88, 185 *et seq*

Handbook of Appropriate Technology 48
Harrison, P. 133
Harrows 77, 78
Hart, Mrs. Judith 181
'Haves' 23
'Have nots' 23
Hayter, T. 28
Heat 190 *et seq*
Heat engine 123
Herbicides 56, 68, 75
High head water lifting 94
Hoe, failure to recognise picture 50, 52
Hookworm 88, 185 *et seq*
Horsepower 190 *et seq*
Hospital equipment, Low cost 13–14, 136
Housing, Low cost 11, 144; traditional 136–7
Hudson Institute 35
Humphrey, H. A. 105
Humphrey pump 105–8, 123, 157, 177
Hyderabad 3
Hydraulic engine 120
Hydraulic ram 104
Hydro power 116
Hydrostatic pump 108

Ibadan, Nigeria 172
Ideas, sources of 45
Illiteracy 21, 192 *et seq*
Impeller pump 98
India 3, 4, 18–24, 30, 42, 44, 48, 63, 68, 115, 136, 155
Indicators, non monetary 19 *et seq*
Industrial output 33
Industry, Small, Problems of 146
Infant mortality 20, 133
Information storage and communication 46
Insect vectors 86, 185 *et seq*
Insecticides 56, 68, 75
Intermediate Technology, I.T. v, vii *et seq*, 3, 40 *et seq*
Intermediate Technology Development Group, I.T.D.G. v, vii, 4, 47, 156, 174, 175
I.T.D.G. panels 46, 134
I.T.D.G. publications dept. 46–7
Internal combustion engine 103, 123

International Bank for Reconstruction and Development, I.B.R.D. 198
International Company Conspiracy 22
International Development Association, I.D.A. 198
International Finance Corporation, I.F.C. 198
International Labour Office, I.L.O. 168
International Maize and Wheat Improvement Centre, C.I.M.M.Y.T. 68
International Monetary Fund 198
International Rice Research Institute, I.R.R.I. 68
Inventors 44
I.R.R.I. bellows pump 100
Irrigation 68, 81, 91
Irrigation trickle 91

Jali 140
Jamaica 167
Japan 113
Joad, C. E. M. 42
Johnson, Dr. S. 146

Kahn, H. 35
Kaldor, N. 28
Kenaf stripper 80
Kenya 18, 20–3, 44, 156, 158, 162
Kerala 140
Khandsari plant 155
Kilowatt 190 *et seq*
King, A. 32
Kitchen gardens 11
Kuby, T. 14
Kumasi, Ghana 5, 44, 150, 152, 175
Kwashiorkor 62, 84
Kwashiorkor patient 83

Labour peaks in agriculture 70
Lahore 5
Land availability 67, 70
Land tenure 6, 70
Lateral thinking 45
Lathe, Simple 122, 125; capstan 157
Latin America 67, 69
Latrine design 109–10
Law of inverse difficulty 177
Leguminous plants 56
Lemmings 32
Levellers 75–6
Lewis, W. A. 27
Life expectancy 20
Lik-lik Buk 48
Lime 141
Lime kilns 141
'Limits to Growth' 32
Liquid piston engines 120, 123, 127
Literacy 21, 192 *et seq*
Lloyd-Wright, F. 176
Lovins, A. 38
Low cost automation 157; hospital equipment 13–14, 136; housing 11, 144; technology 40; work places 3, 29, 40
Low head water lifting 94
Low income countries 1

McRobie, G.   vii, 4, 47
Maintenance of equipment, Need for   71, 115, 138
Maize   66
Makhijani, A.   42
Malaria   64, 87–8, 185 *et seq*
Malaysia   18, 20–3
Malnutrition   62; Effect on school attendance   159
Malthus, T.   31
Mao, Chairman   4
Marasmus   62
Market economy   23
Marsden, K.   16
Marx, K.   38
Marxist economists   26
Mauritius, Wave power for   116
Meadows, D. H.   32
Mechanisation, Agricultural   69
Medellin, Colombia   76
Medical care   133
Medical equipment   133 *et seq*
Medical facilities   20
Medicine, Traditional   134
Melanesian Council of Churches   48
Mesarovic, M.   35
Metal bending machine   143–4
Methane (bio-gas)   43, 103, 117, 123; fuel for engines   103, 123
Methane generator (digester)   23, 117–18
Mexico   68
Micawber, Mr. Wilkins   32
Middle Eastern Oil States   113
Milan   168
Minerals, human requirements   61; plant requirements   56
Mixed farming   67
Models economic   27
Model, World   32 *et seq*
Molière   vii
Monocropping   67
Muscle power   114, 115, 122; world human   113
Muscle efficiency   58
Museum filling   172
Myrdal, G.   26

National Income, N.I.   2, 181 *et seq*
National Sharecroppers Fund   149
Natural gas   116
New Internationalist Magazine   115
N.P.K.   57
Nigeria   13, 44, 46, 78, 83, 135, 172
Nile   90
Nitrogen   56
Nitrogen fixing   56, 66
Nuclear energy   114, 116
Nurses, Distribution of   133
Nutrition F.A.O./W.H.O. Guidelines   62
Nutrition, Human needs   62

Ocean temperature gradients   116
Oil   116
Oke, O. L.   83

Open University   165
Overseas aid   28, 198
Overseas Development Ministry, O.D.M.   108, 198

Pakistan   44, 48, 68, 78
Panels, I.T.D.G.   46; agriculture   46; water   46; power   46; rural health   46, 134
Paramedics   13, 44, 133, 162
Parkes, M. E.   132
Partnership for Productivity, P.f.P.   156
Passive system design   141
Peru   44, 133
Pestel, E.   35
Petrol engine   103, 123
Ph.D. topics   106, 150, 169
Philippines   24, 68, 100
Phosphorus   57
Photocell (photoelectric converter, photovoltaic converter)   120, 128
Photosynthesis   56; efficiency of   59
Pictorial illiteracy   48
Pirie, N.W.   66
Piston engines   120; liquid piston   123, 127
Piston pump   97–8
Plant   56–7
Plantation crops   66
Plough   77; oxdrawn   78; kabanyolo   79
Population   29: growth   30; limitations   31, 33, 134; structure   30, 134
Pollution   32
Politics   6, 29, 38
Porter, Mrs. Julia   vii, 4
Potassium   57
Power   190 *et seq*; density   116; hydro   116, 128; muscle   114, 115, 122; wind   116, 128; units of measurement   190 *et seq*
Power mechanical shaft   116
Power level   114
Power for lighting   116
Pozzolana (surkhi)   141
Processing crop   114
Project selection   53
Projects, Teaching   170; undergraduate   169
Protein   19, 56, 61, 62
Protein Calorie Malnutrition, P.C.M.   63
Protein, New sources of   66
Protein, Leaf   66, 83
Protein, Single cell, S.C.P.   67
Pumps   95 *et seq*; *see also* air lift, flap valve, humphrey, hydraulic ram, impeller, I.R.R.I. bellows, piston, resonant Afghan joggle, Sofretes, solar
Pump priming   100
Puno, Peru   133
Pyrolysis (destructive distillation)   118

Quality of Life   36–7

Radioactive decay   116
Rainfall   85
Rankine cycle   120
Rao, D. N. S.   152
Rao, D. P.   127

Reading university   15, 46, 105, 130, 168
Reciprocating engine   120
Refrigeration   114
Research and development   171; projects   53, 150, 169
Resonant Afghan joggle pump   102
Resources   31
Ricardo – steam engine   126
Rice   68
River blindness   86, 88, 185 *et seq*
Roads   42
Rockefeller Rural Development University   163
Roof materials   139
Root crops   66
Rostow, W.   26
Rubber   66
Rural electrification   42
Rural population   3

Sahel   24
Sandals, Plastic, Manufacture of   16
Samuelson, P. A.   181
Sanitation   110
Saponins   185 *et seq*
Savery, Capt. T.   127
Saunders, R.   161
Savings, relation to growth   27
Scabies   87, 185 *et seq*
School building   162; equipment   166, 167; introduction to A.T. in syllabus   166, 167
School-leaver problem   158
Schistosomiasis (bilharzia)   64, 87–8, 89, 185 *et seq*
Schumacher, E. F.   ix, 1, 3, 4, 5, 14, 29, 38, 40
Scraper   75–6
Seeder   79
Seeder Jab   80
Seeding   75, 79
Self help   3, 6
S.E.N.A. Colombia   163
Sewage   110
Shells, for tube wells   94
S.I. System of Units   179
Singer, H.   6
Sleeping sickness   87–8, 185 *et seq*
Slipping rope winch   94
Small is Beautiful   vii, 3
Smith, Adam   38
Socially Appropriate Technology   40
Sofretes pump   108, 127
Solar energy   113, 114, 119; cooker   43, 122, 124, 167; concentrator   121; converter   120; crop dryer   53, 82; flat plate collector   120–2; powered pump   108; 127
Soft technology   40
South America   65–6
Soviet Russia   4
Spare parts · 115
Spark ignition engine   123
Spider glue   152
Spraying, Crop   81
Sri Lanka   44, 68, 155
Stabilised soil   139

'Stages of growth' theory   26
Steam engine   120, 126
Steelworks   42
Stirling   120, 127; *see also* Fluidyne
Storage crop losses   69
Storage water   108
Subsistence economy   6
Suction   98
Sudan   113, 130, 134, 144–5
Sugar   66
'Sugar Babies'   62
Sugar processing   155
Surkhi, *see* pozzolana
Symbiosis   57
Swenson, R.   145
Swift, Dean   55

Taiwan   68
Tabag   134
Take off, Economic   26
Tanzania   13, 134, 145
Technology Consultancy Centre, T.C.C. Kumasi   44, 150, 152
Technology, Dissemination of   48
'Technologies for Small Industries in Rural Areas'   3
Technological fix   35
Technoserve Inc.   155
Tema Low Cost Housing   11–12
Terracing   76–7
Tett, C.   158
Thailand   44
Thermionic converter   120, 127
Thermocouple   (Thermoelectric , converter)   120, 127
Third World   1
Three dimensional sketches   50
Thring, M. W.   36
Tidal energy   116
Tinker, J.   86
Todaro, M. P.   27
Tonne of coal equivalent. t.c.e.   111
T.O.O.L., Holland   5
Toolbar   78
'Tools for Agriculture'   78
'Tools for Progress'   47
Total head   94
Trachoma   87, 185 *et seq*
Tractors   69; Two wheeled   148
Traditional practices   45
Transport   144
Trophisms   59
Tsetse fly   50, 88, 185 *et seq*
Tube wells, *see* bore holes
Turkey   68
Typhoid   88, 185 *et seq*

Underdeveloped countries   1
Underdevelopment Characteristics   28
Undernutrition   62
Unemployment   27, 28; urban   27; rural   3, 68
United Kingdon   18, 20–3
United   Nations   Children's   Fund, U.N.I.C.E.F.   160, 198

220 *Appropriate Technology*

United Nations Development Programme, U.N.D.P. 198
United Nations Education, Scientific and Cultural Organisation, U.N.E.S.C.O: 159, 198
U.S.A. 18–23, 113
University Liaison Unit, I.L.U. 46
University of the West Indies, U.W.I. 167
Urban drift 3

Vapour cycle engine 120, 127
Vaughan, B. 43
Vegetarianism 61
Vicious circle argument 26
Victorian engineering 45
Vietnam South 148
Village Polytechnics, Kenya 162–4
Village Technology Handbook, V.I.T.A. 48
Vitamins 61
Vitamin content of some foods 63
Vitamin deficiency 63
Volta Dam 90
Voluntary Service Overseas, V.S.O. 176
Volunteers in Technical Assistance, V.I.T.A. 5, 76

Wanochi ox cart 143–4
Wall materials 138
Waste disposal 110
Water and Health 86, 185 *et seq*
Water based diseases 87–8, 185 *et seq*
Water borne diseases 87–8, 185 *et seq*
Water, daily per capita use 85; lifting 92; hammer 104; human daily needs 61; large scale supplies 90; related diseases 86, 88, 185 *et seq*; related insect vectors 87–8, 185 *et seq*; rural supplies 90; seal 110; storage 108; turbine 128–9; treatment 87; washed diseases 87, 185 *et seq*; wheel 120, 128
'Water for the Thousand Millions' 90

Watt 190 *et seq*
Watt, J. 190 *et seq*
Wave energy 116
Wealth spectrum 19
Weeder, mulching 81
Weeding 71, 81
Weir, A. 122
Wells 92; siting 87, 93, 110
West Africa 69, 70, 71, 166
Wheat 68
'Whole Earth Catalogue' 46
Wilson, S. S. 122
Windows 140–1
Wind energy 65, 128 *et seq*
Windmill 128 *et seq*; Cretan 7–8, 44, 103; electricity generation by 129; horizontal axis 129–30; I.T.D.G. 103, 131–2, 156; pumping 103, 129; vertical axis 129–30
Windmill, maintenance and lack of 132
Wijewardine, R. 70, 71, 128
Winner Engineering, Bangkok 44, 148
Wood as a structural material 55
Work and energy 190 *et seq*
Work places, Low cost 3, 29, 40, 146
World Bank 198
World energy use 111; per capita, energy and G.N.P. 111–12
World food production 31, 65
World Health Organisation, W.H.O., estimates of water related diseases 86

Xerophthalmia 63

Yellow fever 88, 185 *et seq*
Young's modulus 56

Zambia 14, 73
Zapp, G. 102
Zaria, Nigeria 13

T49.5 .D86 1979 c.1
Dunn, Peter, 192?-
Appropriate technology : techn
3 9310 00035102 1
GOSHEN COLLEGE-GOOD LIBRARY